Springer Theses

Recognizing Outstanding Ph.D. Research

For further volumes:
http://www.springer.com/series/8790

Aims and Scope

The series "Springer Theses" brings together a selection of the very best Ph.D. theses from around the world and across the physical sciences. Nominated and endorsed by two recognized specialists, each published volume has been selected for its scientific excellence and the high impact of its contents for the pertinent field of research. For greater accessibility to non-specialists, the published versions include an extended introduction, as well as a foreword by the student's supervisor explaining the special relevance of the work for the field. As a whole, the series will provide a valuable resource both for newcomers to the research fields described, and for other scientists seeking detailed background information on special questions. Finally, it provides an accredited documentation of the valuable contributions made by today's younger generation of scientists.

Theses are accepted into the series by invited nomination only and must fulfill all of the following criteria

- They must be written in good English
- The topic of should fall within the confines of Chemistry, Physics and related interdisciplinary fields such as Materials, Nanoscience, Chemical Engineering, Complex Systems and Biophysics.
- The work reported in the thesis must represent a significant scientific advance.
- If the thesis includes previously published material, permission to reproduce this must be gained from the respective copyright holder.
- They must have been examined and passed during the 12 months prior to nomination.
- Each thesis should include a foreword by the supervisor outlining the significance of its content.
- The theses should have a clearly defined structure including an introduction accessible to scientists not expert in that particular field.

Alex Mutig

High Speed VCSELs for Optical Interconnects

Doctoral Thesis accepted by
Technical University of Berlin, Germany

Author
Dr. Alex Mutig
Institute of Solid State Physics
Technical University of Berlin
Hardenbergstr. 36
10623 Berlin
Germany
e-mail: mutig@sol.physik.tuberlin.de

Supervisor
Prof. Dieter Bimberg
Institute of Solid State Physics
Technical University of Berlin
Hardenbergstr. 36
10623 Berlin
Germany

ISSN 2190-5053 e-ISSN 2190-5061

ISBN 978-3-642-16569-6 e-ISBN 978-3-642-16570-2

DOI 10.1007/978-3-642-16570-2

Springer Heidelberg Dordrecht London New York

Cover design: Cover design: eStudio Calamar, Berlin/Figueres

Printed on acid-free paper

Springer is part of Springer Science+Business Media (www.springer.com)

To my grandfather Omar Baizaurov

Supervisor's Foreword

The last three decades of the twentieth century witnessed two major, but entirely different, technological developments. Together these have transformed—and will continue to transform—virtually all of the many and varied societies on our planet: In Europe and in China, in the USA and in India, in Islamic as well as in Buddhist countries.

Storage of information, whether data or pictures, in memories and the processing of this information in computers are the first of these developments. The importance of the invention of the electronic memory is equal to that of the invention of the printing press by Gutenberg, which laid the foundation for general education, in which reading and writing could be taught to every child and not just a few selected wise men and monks. Soon we will be able to carry the contents of the Library of Congress with us on a memory stick. The inventions of the transistor and of integrated circuits, both of which were honored by Nobel prizes, were decisive milestones for the still continuing development of memories. Electrons and charges are used to store the bits of information.

Communication, transmission of information and data, any time, anywhere, almost any quantity, at a speed limited only by the speed of light is the second major development. Identical information is accessible at the same time to everybody all over the world, and again not the privilege of a few anymore, as long as each human being knows how to read and write and his/her access to information is not controlled. The invention of the telegraph, utilizing properties of long wavelength electromagnetic waves, was at the advent of the modern communication society. The invention and continuing development of the telephone, based on electrical signals carried by copper-based cables, was the second step. Today again electromagnetic waves are the carriers of information, but now at wavelengths many orders of magnitude shorter than a hundred years ago, in the form of photons transmitted on glass fibers,which are replacing electric signals on copper cables step-by-step. The inventions of the edge-emitting semiconductor laser and of low loss transparent fibers, again honored by Nobel prizes, are the milestones of photonic communication technologies.

Today, more than three decades have passed since photons made their debut carrying information on glass cables across distances of hundreds and then thousands of kilometers. High frequency modulation, wavelength multiplexing, higher order modulation formats, etc. now enable these systems to carry petabits of information from continent to continent, forming the basis of the Internet and the large variety of web-based services. At shorter distances, optical communication across tens and hundreds of kilometers enabled online banking and video on demand, soon complemented by streaming. Fiber to the home and fiber in the building are the present extensions of optical networks to shorter and shorter distances, presently still dominated by copper cables. Human-centered communication was in the focus of these first three decades of system developments.

Machine-centered communication at much shorter distances, from meters to millimeters or, in the more remote future, to micrometers, is the next area being conquered by optics. The first steps have already been made. Computer, memory and communication technologies are merging. Modern computers are limited in their performance by the speed of data input from and output to memories. IBM's latest supercomputers employ therefore millions of semiconductor lasers serving optical links.

The data storage capacity of flash memories, a typical consumer product, is doubling every 18 months. The data transmission speed presently controlled by standardized copperbased connectors, being part of the USB 2.0 standard, is much too slow for fast transfer of large amounts of data, and presents a performance bottleneck preventing full use of the next generations of flash. The USB 3.0 standard, being currently introduced, will have dual electrical and optical interfaces and the following USB 4.0 standard will be solely optical.

Optical board-to-board and chip-to-chip interconnects in computers is the other challenging subject of research and development in the near future.

A completely different type of laser, invented only about two decades after the edgeemitting laser, the vertical cavity emitting laser (VCSEL), is employed for machine-centered communication. Its footprint is typically two orders of magnitude smaller than that of an edge emitter, which means that an array of 100 of such lasers takes as much surface as one edge emitter. VCSELs can be tested on-wafer, have a symmetric beam profile, meaning that the focusing optics are inexpensive, deliver output power in the range of milliwatts and can be much more temperature stable than typical edge emitters in those properties decisive for system applications. Their cost of production can be scaled down under future mass production conditions to values close to those of LEDS. Such lasers were initially ignored for applications because of their very large threshold current density and limited reliability and power conversion efficiency until the mid 1990s. Then the oxide aperture technology was invented, solving all these problems at once, at least for GaAs-based devices. Later on different solutions were developed to solve these problems for InP-based VCSELs.

The wavelengths of 850 and 980 nm presently employed for machine-centered communication systems are much shorter than those for human-centered communication,which are 1310 and 1550 nm, but the demands on modulation capabilities of the devices are identical: the faster the better. In addition the demands on stability of all properties across a large temperature range up to 85 or 120°C are enormous. Such short-wavelength VCSELs are based on GaAs technology. Until recently, most of their commercial producers argued, based on their practical experience, that the maximum modulation frequencies of GaAs-based VCSELs are limited at room temperature to 10 GHz, allowing data rates of about 15 Gbit/s at bit error rates of 10^{-12}. The cutoff frequency is controlled by intrinsic properties of the gain medium, parasitics and thermal effects. However, three-dimensional numerical models of the dynamics of VCSELs, solving self-consistently the Schrödinger, Maxwell and Helmholtz equations, which would allow a reasonable prediction of the maximum modulation rate, presently do not exist: The challenges involved in solving these equations across 12 orders of magnitude in time and 6 orders of magnitude in space are large.

Based on one decade of experience of VCSEL development in my research group, Alex Mutig, in his diploma work and the present Ph.D. thesis, has investigated and challenged the limits of high speed/bit rate operation of these devices using a semiempirical approach. He investigated in great detail all the different relevant properties and limits of the various types of VCSELs he designed, also as a function of temperature, and traced them back to the three groups of "fundamental" limits, using separate modeling approaches. Astoundingly he discovered that these limits can be pushed to values almost a factor of two larger than previously thought if one modifies both the vertical structure—including the gain medium—and the processing technology. In addition, he demonstrated that the dynamic properties of high speed VCSELs can be maintained up to the high temperatures demanded by system manufacturers. Following his systematic approach, further extensions to larger bandwidth and higher temperature seem to be possible.

This book, based on his Ph.D. thesis, discusses therefore all relevant design issues, the technologies required and high speed/data rate measurement techniques up to the 40 Gbit/s reached for short-wavelength VCSELs in a way understandable for nonspecialists. I wish you an interesting and enlightening time reading the book.

Berlin, August 2010 Prof. Dieter Bimberg

Acknowledgments

First of all I would like to thank my advisor Prof. Dr. Dieter Bimberg for the years of immense support during my PhD work, for defining the very exciting and future oriented research topic, for building and leading the very strong research group, including many very experienced guest researchers, and for creating perfect conditions for efficient research and development, including among other complete growth, device fabrication and device characterization lines. Without all of these preconditions this work would be impossible.

I would in particular like to thank Dr. James Lott, who has shared with me his immense experience in the VCSEL field and his physical understanding of the laser problems, especially at high speed. I appreciate his great support during the simulations of VCSEL structures and not least the enjoyable communication over the years, which has among other improved my English.

My special thanks goes to Prof. Dr. Nikolai Ledentsov for his deep understanding of the laser physics, for his perfect analytical abilities, for his great ideas and not least for his never ending optimism and for very enjoyable and fruitful discussions. I greatly appreciate Prof. Dr. Vladimir Haisler for bringing me a deeper understanding of the physical processes in semiconductor lasers and for the great support over the years. Special thanks to Dr. Vitaly Shchukin for explanations of the VCSEL physics. I would like specially to thank Prof. Dr. Shun Lien Chuang for the very interesting and informative discussions not only about the physics of photonic devices.

Of course, I want to thank the VCSEL group as a whole and everyone individually. The two Philips, Philip Moser and Philip Wolf, have made a great job building the best oxidation furnace a university can have. They have greatly supported me, especially in the last two years of my dissertation work, making CW and complicated high frequency measurements and fabricating directly and indirectly modulated VCSELs. I greatly thank Dr. Friedhelm Hopfer, who is since more than a year in Singapore, for advising and teaching me in the first years of my work at TUB. His perfect accuracy during the VCSEL fabrication process in the clean room has impressed me and affected all my future work. I appreciate Dr. Werner Hofmann, who has joined our VCSEL group this February, for exciting ideas for the future VCSEL designs and for very fruitful and very enjoyable discussions of the VCSEL physics.

I am especially grateful to our Russian colleagues from the Saint-Petersburg Ioffe Institute, in particular Dr. Sergey Blokhin, Dr. Leonid Karachinsky, Alexey Nadtochiy, Dr. Innokenty Novikov and Alexey Payusov for their immense support during the laser characterization over the years and very fruitful and enjoyable discussions. They have measured day and night, also at weekends, to achieve the worldwide best VCSEL results. Without this decisive collaboration the present work would be impossible.

I would like to thank the whole research group of Prof. Dr. D. Bimberg for the permanent immense support, both theoretically and experimentally. Special thanks to Gerrit Fiol and his student Dejan Arsenijevic, who have greatly supported us during the high speed laser characterization.

Special thanks to all of our colleagues from the Fraunhofer-Institut Heinrich-Hertz-Institut für Nachrichtentechnik in Berlin (HHI) for their support with their immense experience in device fabrication and high speed characterization. I am especially grateful to my advisor at HHI Dr. Harald Künzel for his permanent help and support.

I am very thankful to all my relatives and friends, who have supported me during these years.

Of course, without the love of the family members nobody can live and work. I am endless thankful to my two mothers, Rimma Mutik and her sister Emma Bauzaurova, for their everlasting mother's love. Without them I would not be in this world, and of course could not make this research.

Special thanks go to my brother Dr. Kerim Mutig for his brother love and the never ending help and support in all situations. I would like to thank also his wife Tanja Mutig and his son Damir Mutig for the happiness to be uncle.

I would like to thank my father Valentin Mutig, since he has also decisively contributed to the fact that I am existing, and has spent practically all his time with me and my brother in my childhood, making us strong and teaching us for the rest of the life.

I am endless thankful to my grandfathers Omar Baizaurov and Josef Mutig and to my grandmother Maria Mutig for their love and immense support. They have survived during the most terrible times not only of the twentieth century but also of the whole human history until now. I wish I would have a smallest part of their wisdom and cleverness, but first of all of their kind-heartedness. Until now I wonder about their incredible predictive efficiency. I can remember how my grandfather Omar, who was born in 1910, as Zar has ruled Russia and nothing of the modern things we all know existed, has many times told to me, as he was more than 80 years old, that me and my brother should learn to speak English and to work with a personal computer. I wish I could tell my grandchildren the right things about the future world, which will exist in 2060.

Finally, I thank the most important people in my life: to my wife Julia Mutig and to our first son Timur Mutig. They give me the happiness, the love and the meaning to all my life. I am endless grateful for their patience and for their immense help and support. Finally, I want to thank the smallest man I know — our second son Eldar Mutig, whose love helps me in all my life.

Contents

Chapter 1
Introduction

1.1 Introduction

The progress of human society continues at a permanently increasing tempo. Acceleration of both individual and social human development creates new challenges and requires ever faster reactions to the rapidly changing conditions. While at the beginning of the human era progress was caused mainly by evolutionary changes of the human body, later progress in the intellectual sphere began, and is nowadays the main driving force of the development of the human society, as illustrated by Fig. 1.1 [1]. Communication and information exchange between individual members of society play here a decisive role. To know not only how to do something but also when and where became an indispensable precondition for the success of an individual as well as of the whole society. Starting from simple verbal information exchange, communication has developed through the invention of script, books, printed mass media, radio, telephones, television and, more recently, computers and the Internet to a vital basis of our society. All of industrial, technological, scientific, cultural and social progress would be impossible without efficient communication.

Efficient communication requires advanced technologies for fast data transfer from one place to another. While in the past fire, smoke, doves and horses were used to send a message, today modern data transmission lines based on copper wires and optical fibers or even wireless information exchange over radio, television and satellites are utilized, so that signals are transmitted by electrons and electromagnetic waves. In the following, an overview of modern communication technologies will be presented, special attention being paid to the ever increasing role of optical data transmission lines and particularly to vertical cavity surface emitting lasers (VCSELs) [2–6], as one of the most promising laser light sources for the future optical interconnects [7–10].

A. Mutig, *High Speed VCSELs for Optical Interconnects*, Springer Theses,

DOI: 10.1007/978-3-642-16570-2_1,

Fig. 1.1 Human progress: from the evolution of the body to the evolution of the mind

1.2 Optics in Telecom and Datacom and the Role of VCSELs

The length of a data transmission line has a strong impact on the quality of the transmitted signal and determines the absolute values of the important physical quantities, e.g. electrical resistance, optical absorption, optical dispersion, and overall losses. Therefore it appears logical to classify data transmission lines with respect to their length. The longest communication lines span distances of many thousand kilometers, for example the intercontinental fiber-based links between North America and Europe, Asia and Australia, etc. The shortest interconnects could be only several micrometers or even shorter in length, for example within a microprocessor chip in a personal computer.

Since computers, both personal computers and high performance computers (HPC), are nowadays the main tools for the information processing, it makes also sense to distinguish between data transmission lines outside computer and inside computer. In Fig. 1.2 the hierarchy of the data links outside computer is presented. On the bottom arrow the today's role of the optical technologies on the corresponding hierarchy level together with the approximate starting date of their application in the past are shown (based on [11]).

Depending on the transmission distance one can distinguishes between the global area network (GAN), wide area network (WAN), metropolitan area network (MAN) and local area network (LAN). Historically one speaks by longer distances, like in the case of the GAN, WAN and MAN, about telecommunication (Telecom), while by shorter distances, like within a LAN or shorter, it is common to use the term data communication (Datacom). Transmission distances within the GAN can reach many thousand kilometers, for example in intercontinental links, while a WAN commonly connects different cities of one or several countries with typical distances of several tens to thousands of kilometers. The area of a MAN is mostly limited to the area of a city and a LAN commonly transfers data between several computers within one building or a building complex.

As can be obtained from Fig. 1.2, optical technologies are ubiquitous in all types of the networks since the end of the last century. While at longer distances

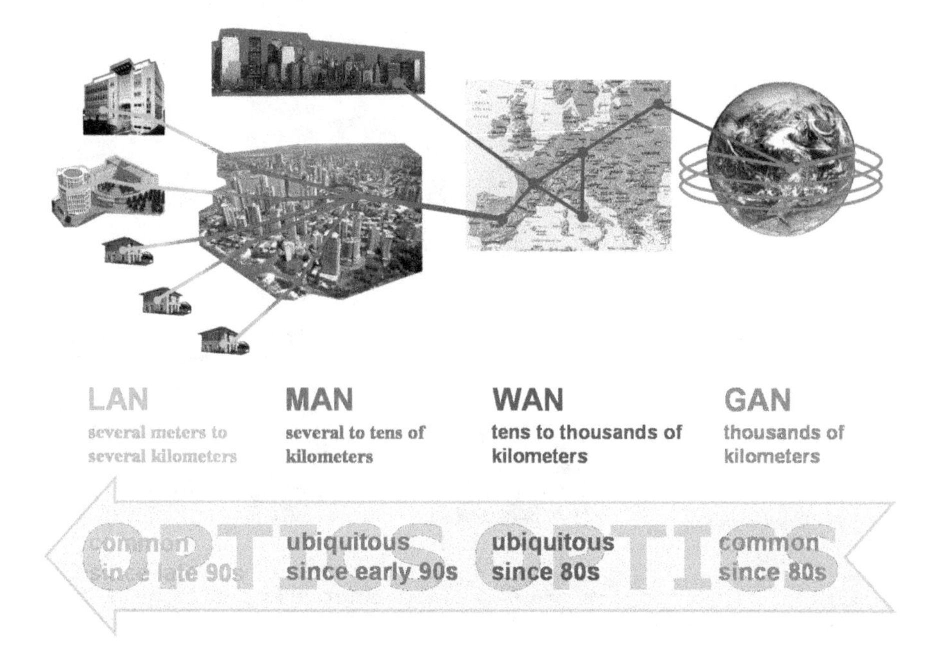

Fig. 1.2 Network hierarchy for the data transmission lines outside computer with a short description of the role and the approximate starting date of the usage of the corresponding optical technologies

edge emitting lasers at wavelengths around 1550 and 1310 nm, at which respectively minimum of absorption and dispersion in the optical fibers occur, are utilized, for the LAN applications oxide-confined VCSELs [12, 13] emitting around 850 nm have established themselves as very reliable and cost-effective laser light sources. The success of the VCSELs is determined by their unique physical properties, leading to decisive advantages for short reach data communication as compared to the edge emitting lasers, which are among other [14]:

- a near circular output beam with a small divergence angle, determined by the circular shape of the VCSEL mesa and thus the near circular shape of the oxide aperture;
- stable modes and emission spectrum, determined by the stability of the physical properties of the VCSEL cavity;
- low threshold current and low power consumption, determined by the small volume of the active region;
- large differential efficiency, determined by the controllable internal and mirror losses;
- high temperature stability at milliwatts of peak output power, determined by the lower temperature shift of the cavity resonance wavelength as compared to the larger shift of the gain peak wavelength, caused by the larger shift of the band gap energy of the active material;

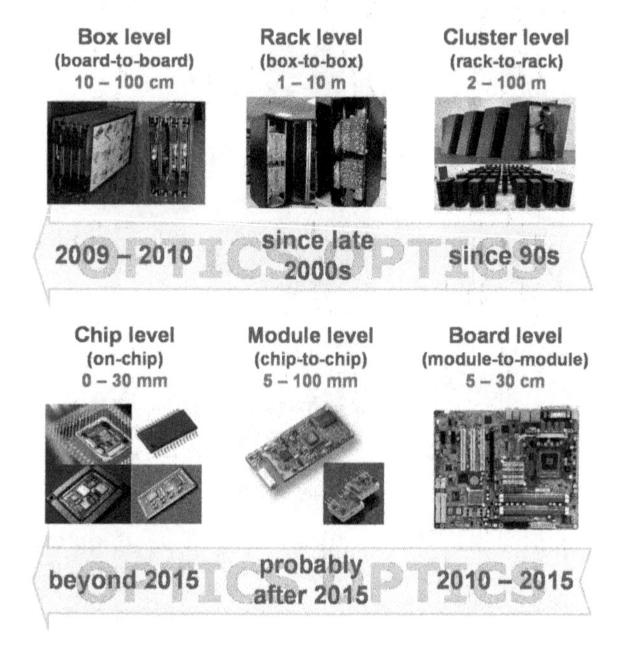

Fig. 1.3 Different levels of interconnects inside computer and approximate date of the entering of the optics into the corresponding market segment

- planar processing and on wafer characterization, enabling inexpensive production and testing of VCSELs and 2-dimensional VCSEL arrays;
- high reliability and easy packaging.

Oxide-confined VCSELs emitting around 850 nm are commercially available since roughly a decade. Many millions of VCSELs are fabricated yearly with very mature, robust, cost-effective large-scale mass production processes by many well established companies [15–17]. Today 850 nm oxide-confined VCSELs operating at bit rates of up to 10 Gbit/s are commercially available on the market.

To bring the signal to a computer is only one important part of the information transfer process. The second, equally important part is to provide technologies for data exchange between different components within a computer. Again, for this task one can distinguish between different levels of interconnects inside computer, depending on the transmission distance, as shown in Fig. 1.3.

Together with the different levels of interconnects inside computer also approximate date in the past or predicted date for the future for the entering of optics into corresponding market segment is shown (based on [18]). The distances of interconnects inside computer could vary from several hundred meters for the rack-to-rack connections in a supercomputer down to only several millimeters or even several micrometers for the on-chip interconnects. Nowadays optics is common on the cluster level for the rack-to-rack interconnects and on the rack level for the box-to-box interconnects. Currently optical technologies are coming to the box level for the board-to-board interconnects with data transmission distances of several tens of centimeters. For this purpose active optical cables (AOC) mainly based on short wavelength VCSELs are widely used, which have decisive

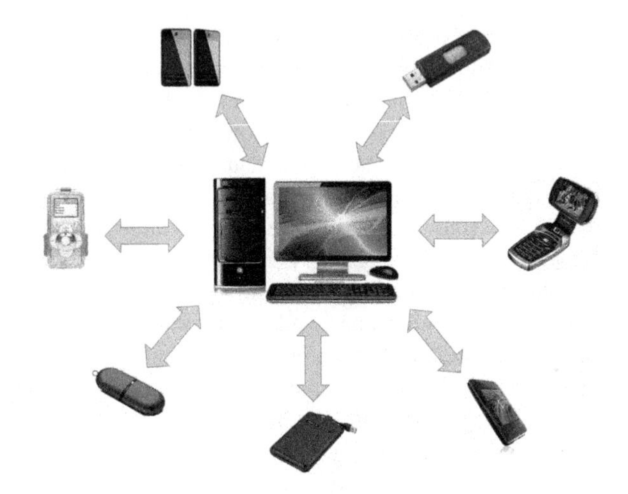

Fig. 1.4 Personal area network (PAN): a personal computer with portable devices like USB-sticks, mp3-players, mobile phones, etc. with typical data transmission distances of up to several meters

advantages as compared to the copper-based cables, which are among other longer transmission distances, less weight, less volume, smaller bend radius and lower bit error rates (BER) [19]. Since no wavelength was standardized for this kind of applications until now, different companies use VCSELs emitting at different wavelengths, commonly around 850, 980 or 1100 nm. The market of the AOC is rapidly growing and is predicted to exceed 2.4 billions US dollars in the year 2014 with approximately 48 millions cables installed [20]. Compared to the predictions for the year 2010 this would be a growth by more than one order of magnitude, opening unique perspectives for the significant growth of the market for high speed short wavelength VCSELs. A noticeable part of the growth of the AOC market and thus of the VCSEL market would be caused by the expansion to applications based on the universal serial bus (USB) and high definition multimedia interface (HDMI) standards. Since a portable device could be directly connected to a computer, such types of interconnects belong with respect to their length to somewhere between interconnects outside computer and interconnects inside computer, and can be considered as a part of the personal area network (PAN). A PAN connects a computer to portable devices like USB-sticks, mobile phones, etc., as demonstrated by Fig. 1.4, and has typical data transmission distances below 10 m. Also future optical interconnects for data transfer over distances up to several meters between a computer and portable devices, like for example those based on the Light Peak technology [21] recently announced by Intel, could be considered as a part of a PAN and will definitively lead to the noticeable growth of the VCSEL market.

Penetration of the optics into ever shorter interconnects will continue in the near future, as can be obtained from Fig. 1.3. Just in the next few years it is expected that optical interconnects will expand to the board level for the module-to-module data transfer [22]. After approximately 2015 optics will come also to the module level for the chip-to-chip and to the chip level for the on-chip data transmission. The strong attenuation and distortion of the high frequency signal in the copper-based links at distances of 1 m and longer, which are typical for the data transmission on the cluster, rack and box levels and also for the LAN and other types of networks, and at frequencies above 10 GHz lead to the unavoidable exchange of the copper-based interconnects by the optics in these application areas. At shorter distances the physical copper limitations caused by the high density of interconnects, among other electromagnetic interference (EMI), crosstalk and high power consumption, are the main driving forces for the usage of optics there [23–27]. However there are also additional requirements for the electro-optical components for applications at shorter transmission distances. The main required property additionally to the high speed is the high temperature stability, especially for the laser light sources [28], since lasers should be placed directly near the data source on-board, inside the module or even directly on-chip, where temperatures could achieve values of 85°C or even higher.

The persistent pressure on the market leads to significant investment of recourses into the research and development of high speed optical components capable to replace the copper-based links in the future interconnects. Both industrial companies and academic organization are currently working on these topics. While for the optical data transmission lines outside computer (Fig. 1.2) general concepts are relatively well established and the main effort is to improve the existing performance, for optical interconnects inside computer (Fig. 1.3) the concepts themselves are also under investigation in different projects like Terabus from IBM [29, 30] or Multiwavelength Assemblies for Ubiquitous Interconnects (MAUI) from Intel [31]. The ultimate goal for the optical interconnects outside computer is to bring data with a very high speed using optical technologies directly to the home, the so-called fiber-to-the-home (FTTH) concept [32]. The ultimate goal for the optics inside computer is to transfer data by light at all levels, even at the very short distances on-chip. As distances become shorter and the density of interconnects and the number of the transmission lines grow, the fabrication costs of a single laser start to play an ever increasing role in the total costs. VCSELs, as very inexpensive low power reliable laser light sources, benefit from this trend decisively. Nowadays oxide-confined GaAs-based VCSELs dominate the Datacom market.

The cheaper alternative to the GaAs-based VCSELs would be Si-based optical interconnects, which would strongly benefit from the well established inexpensive silicon fabrication technology [33]. However the main obstacle of the silicon photonics is the nature of the silicon, which is a semiconductor with an indirect band gap, making lasing in the silicon material practically impossible. Different approaches have been proposed to overcome this problem and significant results have been achieved on the field of silicon photonics in the past, among other successful demonstration of the distributed feedback silicon evanescent laser (DFB-SEL) with the active region based on III–V materials [34–36], AlGaInAs-silicon evanescent racetrack lasers and photodetectors [37], mode-locked silicon evanescent lasers operating up to frequencies of 40 GHz [38], AlGaInAs-silicon Fabry–Perot (FP) lasers [39], hybrid AlGaInAs-silicon evanescent amplifiers [40], Ge-on-SOI waveguide photodetectors operating at bit rates of up to 40 Gbit/s [41] and silicon-based modulators operating at bit rates of up to 40 Gbit/s [42]. These approaches however utilize novel concepts, which require maturity proof to show their suitability for the commercial large-scale cheap mass production and reliable operation over years. Oxide-confined GaAs-based VCSELs have compared to these novel Si-based concepts the major advantage of the market presence since roughly a decade and have been proven to be very reliable and inexpensive laser light sources, while having similar high speed characteristics, as we will see in the next sections.

As can be obtained from the overview of the optics in the Telecom and Datacom applications, optical interconnects are ubiquitous for data transfer outside the computer on the LAN, MAN, WAN and GAN levels. They penetrate unstoppable to ever shorter distances and will reach in five to ten years the chip level, The major role as inexpensive, low power and reliable laser light sources on all levels of optical interconnects starting from LAN down to the cluster, rack and box levels belongs to the GaAs-based oxide-confined VCSELs. Future development and expansion of the optical interconnects to the board, module and chip levels will be definitely based on the short wavelength VCSELs, at least until potential novel revolutionary concepts will be investigated and ready to enter the market. Consequently, the demand for high speed VCSELs will continue to grow, driving the research and development in this area to new records and bringing new VCSEL-based products to the market.

1.3 Moore's Law in Datacom

After we have investigated the role of the optical interconnects in the overall data transmission process and especially the role of the VCSELs as the major laser light sources for Datacom application both outside and inside the computer today and in the future, the next question is: what is the main goal of the research and development on the area of high speed VCSELs for optical interconnects? To answer this question one can consider the driving forces of the improvement of the performance of interconnects.

More than 40 years ago the great visionary and co-founder of Intel Gordon Moore has introduced an empirical law, which predicts that the number of transistors on a chip will double approximately every two years [43–45]. Nowadays this law is known as the Moore's law, by the name of its inventor. This trend, predicted in the year 1965, holds over decades and is expected to continue also in the future. International Technology Roadmap for Semiconductors (ITRS) predicts in the edition 2007 that the future trends of the cell and logic gate size and of the number of transistors per chip will follow the Moore's law also in the next decade [46], as demonstrated by the Fig. 1.5.

The direct conclusion of the Moore's law is the permanent increase of the computational power of computers, based on chips capable to handle larger and larger amount of information at ever increasing speed. Directly in consent to

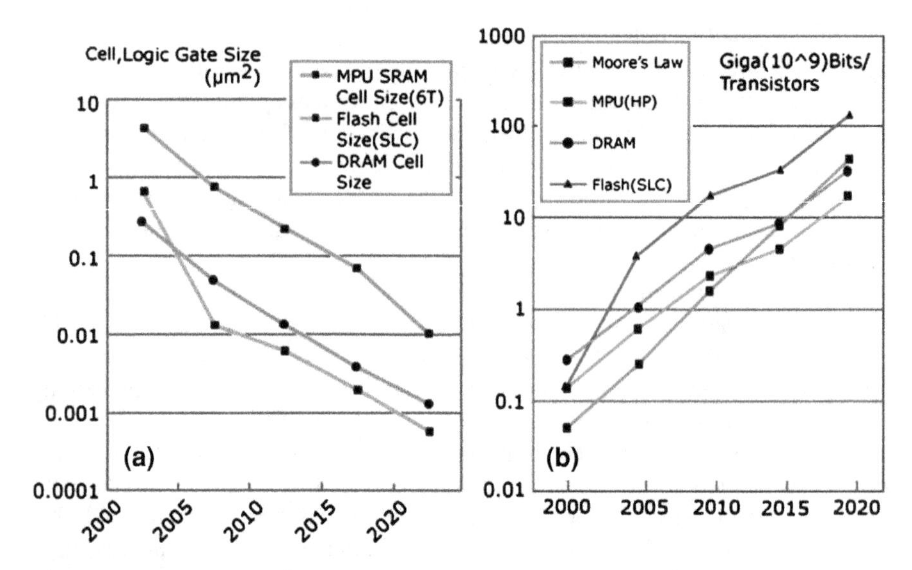

Fig. 1.5 2007 ITRS trends: product function size trends—cell size and logic gate size (**a**) and product technology trends—functions per chip (**b**)

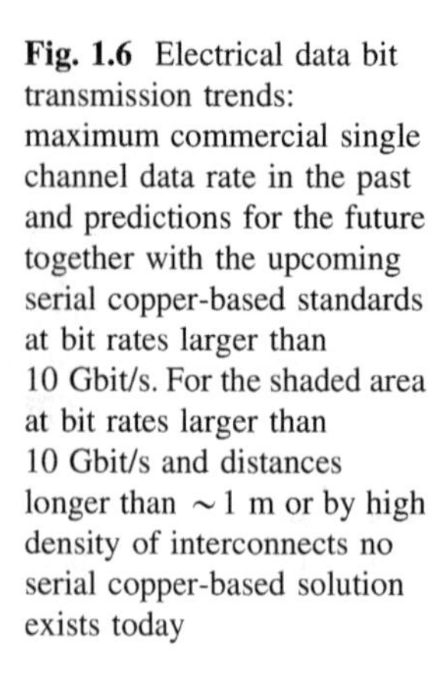
Fig. 1.6 Electrical data bit transmission trends: maximum commercial single channel data rate in the past and predictions for the future together with the upcoming serial copper-based standards at bit rates larger than 10 Gbit/s. For the shaded area at bit rates larger than 10 Gbit/s and distances longer than ~1 m or by high density of interconnects no serial copper-based solution exists today

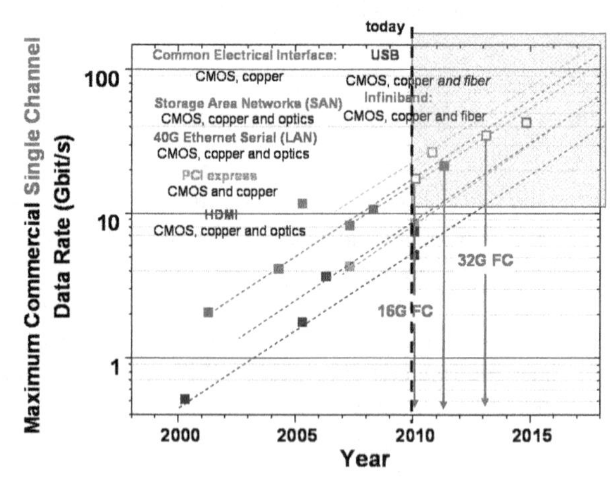

silicon integrated circuit scaling, the forecast for the serial transmission data bit rates standardized for interconnects is a continued exponential increase with time, as shown in Fig. 1.6, where maximum standardized commercial single channel data rate in the past and also predictions for the future are demonstrated for different standards, among other Common Electrical Interface (CEI), HDMI, USB, Ethernet, etc.

All presented standards follow the same trend: the single channel data rate increases exponentially with the time by a factor of approximately four within roughly five years. Today single channel data rates reach the value of 10 Gbit/s. Following the trend, electrical interfaces for serial transmission at speeds beyond 10 Gbit/s are being standardized for a variety of applications including for example (with an expected data-rate): Fibre Channel FC32G (34 Gbit/s), Infiniband (20 Gbit/s), common electrical interface CEI (25–28 Gbit/s), and universal serial bus protocols USB 3.0 (accepted up to 25 Gbit/s). For such high bit rates the maximum copper wire-based link is limited to distances below 1 m at the highest speeds, so that no serial copper-based solution for the shaded area in Fig. 1.6 at bit rates larger 10 Gbit/s for mostly applications exists at the moment. Thus the only chance to fulfill the standards and to reach data rates above 10 Gbit/s at distances longer than $\sim$1 m is the usage of optical interconnects. Because of the large number of required links, the market pressure, enabling only very inexpensive solutions, and the needed low power consumption, defined by the nature of applications, VCSEL-based interconnects are today the only realistic solution.

Thus, one can conclude, that the progress in the silicon integrated circuits, following the Moore's law, immediately drives the progress of the data transmission lines, requiring ever higher bit rates, which can be enabled only by the optical technologies. The bit rate of 10 Gbit/s for the serial transmission is the border, after which copper-based links bump into their physical limitations. Just in the nearest future VCSEL-based optical interconnects operating at bit rates of 20–30 Gbit/s and even higher will be urgently required by the market. Since the maximum bit rate of the present commercially available VCSELs is 10 Gbit/s, the main goal of the VCSEL research and development is to push the bit rate to higher values, simultaneously meeting other important requirements, among other high temperature stability, low power consumption, excellent reliability, etc., and maintaining the established mature inexpensive large-scale mass production VCSEL technology, without introducing new complicated growth or device fabrication steps.

1.4 Recent Progress on High Speed VCSELs

Following the motivation presented in the previous sections, many industrial companies, universities, research centers and institutes have focused their activity on the development of high speed GaAs-based VCSELs emitting at wavelengths around 850, 980 and 1100 nm. This has led to a significant progress on the field of short wavelength VCSELs in the recent years. At different wavelengths different maximum bit rates have been measured, but the overall trend is unmistakable: at all three wavelengths bit rates beyond 30 Gbit/s have been achieved within the last three years and the expectations for new records remain very high. Also at longer wavelengths, for example at 1310 [47] and 1550 nm [48], the trend goes to larger bit rates. The remarkable progress on the field of high-speed VCSELs reflects the immense effort made by the researchers all around the world to increase the bandwidths of the lasers. Many novel concepts, among other photonic crystal VCSELs (PhC-VCSELs) [49, 50], proton implanted holey VCSELs [51] and

VCSELs with novel active regions [52], have been applied in order to increase the modulation bandwidth. In the following an overview about the state-of-the-art high speed VCSEL results will be presented, distinguishing between different emission wavelengths, since important physical properties, like free-carrier absorption, non-radiative recombination, etc., and corresponding challenges are different depending on the laser wavelength. Thereby we will limit our discussion to the shorter wavelengths of 850, 980 and 1100 nm, since these are the wavelengths of the GaAs-based high speed VCSELs important for the Datacom applications.

1.4.1 State-of-the-art of 850 nm VCSELs

The most interesting wavelength from the standardization point of view is the wavelength of 850 nm, since it has been already standardized for LAN and storage area network (SAN) applications and definitely will play an important role in the upcoming future standards. Excellent infrastructure for applications targeting 850 nm exist, including optical fibers optimized to 850 nm and relatively inexpensive GaAs photodetectors operating around 850 nm. The wavelength of 850 nm has also several physical advantages compared to longer wavelength. The major physical advantage is the lower free carrier absorption coefficient of the AlGaAs material as compared to longer wavelengths, since the absorption coefficient increases with the wavelength. Combined with the thinner layers, caused by the shorter wavelength compared to 980 or 1100 nm, this results in the overall lower absorption, enabling higher doping levels and thus lower electrical resistances, and decreasing threshold carrier densities, which is decisive for high speed operation.

Different materials for the active region emitting around 850 nm can be utilized and have been reported, among other GaAs itself [53], InGaAs [54, 55], InAlGaAs [56] and InGaAsP [57]. In the most of commercially available 850 nm VCSELs GaAs quantum wells (QWs) are utilized, which have confirmed their maturity and reliability within the last decade. By adding of In into the active layers compressive strain can be introduced, increasing the differential gain and thus the laser speed. This concept is used in the present state-of-the-art high speed 850 nm VCSELs and also in the devices developed in the scope of this dissertation, enabling to reach error free operation at bit rates as high as 38 Gbit/s with very good perspectives to reach bit rates of 40 Gbit/s and even more.

Already in 2001 oxide-confined VCSELs operating at 20 Gbit/s and emitting around 850 nm have been demonstrated by IBM [58]. It took practically seven years to increase the bit rate of 850 nm VCSELs to 30 Gbit/s. Finally in 2008 the group of Finisar has demonstrated 850 nm VCSELs with GaAs QWs as active region operating at 30 Gbit/s with the maximum bandwidth of 19 GHz [53]. Starting from 2008 the development of high speed 850 nm VCSELs speeded up. Within just two years the maximum bit rate was increased to 38 Gbit/s. The first step was done by the research group of Prof. Dr. Larsson from the Chalmers

University of Technology in Sweden, who has applied compressively strained InGaAs QWs as active medium and successively increased the bit rate from 25 Gbit/s [59] to 28 Gbit/s [60] and further to 32 Gbit/s [54, 55]. One of the major improvements additionally to the compressively strained InGaAs QWs was the introduction of the double oxide aperture, decreasing the parasitic capacitance and increasing the maximum modulation bandwidths, which was as high as 20 GHz. At the elevated temperature of 85°C these VCSELs operated error free at bit rates up to 25 Gbit/s.

Although the bit rate of 32 Gbit/s has been reached, the way to reach 40 Gbit/s still appeared to be long, until the authors have pushed the maximum achieved bit rate of 850 nm VCSELs to 38 Gbit/s [61, 62], bringing the bit rate of 40 Gbit/s within reach. Additionally to the compressively strained InGaAs QWs in the active region and to the double oxide aperture, optimized cavity region together with advanced device design with two mesas and thick dielectric layers was applied to increase the relaxation resonance frequency and to reduce electrical parasitics, both increasing the bandwidth. These lasers are at the moment to the best of our knowledge the worldwide fastest VCSELs operating at 850 nm and the fastest devices of any oxide-confined VCSELs.

1.4.2 State-of-the-art of 980 nm VCSELs

The wavelength of 980 nm has several advantages as compared to 850 nm, in spite of larger free carrier absorption coefficient. The main advantages are deeper barriers, suppressing escape of non-equilibrium carriers and thus improving temperature stability of the gain, the easier possibility to apply strained materials to improve differential gain, the transparency of GaAs at this wavelength, enabling light extraction through the bottom mirror and the substrate, and lower operating voltages, caused by the lower photon energy, which is very important for the low voltage complementary metal oxide semiconductor (CMOS) drivers. The wavelength of 980 nm is nowadays commonly used for pumping Erbium-doped fiber amplifiers (EDFA). However, in view of the advantages described above it becomes also important for VCSEL-based Datacom applications, leading to research and development of high speed and also high temperature stable 980 nm VCSELs and VCSEL arrays.

In the framework of the Terabus program 980 nm VCSELs and VCSEL arrays operating at bit rates of up to 20 Gbit/s were developed by IBM already in the years 2005 and 2006 for the future chip-to-chip optical interconnects [29, 30, 63]. One year later, in 2007, the maximum bit rate for the error free operation of 980 nm VCSELs was drastically increased by the group of Prof. Dr. Coldren from the University of California in Santa Barbara to 35 Gbit/s [64, 65]. They have used tapered oxide aperture with deep oxidation layers and optimized doping profiles, leading to achieved maximum bandwidth of larger than 20 GHz. These are currently the worldwide fastest 980 nm VCSELs.

Although such high bit rate at room temperature was achieved, at elevated temperatures the maximum bit rate was initially limited to 20 Gbit/s at 85°C, demonstrated by our group at the Technical University of Berlin in 2006 [66, 67]. The VCSELs have utilized active layers grown in the submonolayer (SML) growth regime. By further improvements of the device design by the authors within the framework of this dissertation the operation temperature could be increased to 120°C for the high speed VCSEL operation at the bit rate of 20 Gbit/s [68, 69], which is at the moment to the best of our knowledge the highest temperature at which an open eye operation of any VCSEL at any wavelength at 20 Gbit/s have been reported. These were to the best of our knowledge worldwide first VCSELs showing an open eye at 20 Gbit/s at such high temperatures. In the year 2007 Agilent researchers have demonstrated 980 nm VCSELs operating at 25 Gbit/s at 70°C [70]. In the framework of this dissertation 980 nm VCSELs demonstrating open eye diagrams at the bit rate of 25 Gbit/s at temperatures of up to 85°C were realized by the authors, increasing the operation temperature of 980 nm VCSELs at 25 Gbit/s further.

1.4.3 State-of-the-art of 1100 nm VCSELs

The wavelength of 1100 nm provides even more advantages from the thermal point of view because of the even larger possible band gap discontinuities compared to the wavelength of 980 nm. Operation voltages are lower as well. However, the negative trade-off is the higher losses at 1100 nm compared to shorter wavelengths. Nevertheless this is the wavelength at which first VCSELs operating at bit rates as high as 40 Gbit/s have been demonstrated [71, 72]. These were InGaAs-based oxide-confined 1100 nm VCSELs developed by the researcher from NEC.

Already in the year 2006 this group has demonstrated oxide-confined VCSELs emitting at 1100 nm and operating at bit rates of up to 25 Gbit/s [73]. Later also 25 Gbit/s operation of 1100 nm VCSELs at the elevated temperature of 100°C has been presented by the same group [74]. In the year 2007 the NEC group has demonstrated 1100 nm VCSELs based on the buried tunnel junction, which have operated at the bit rate of 30 Gbit/s [75–77]. Finally in the same year the bit rate of 40 Gbit/s using VCSELs based on the buried tunnel junction has been achieved by the NEC group [71, 72]. These are presently the fastest VCSELs of any VCSELs. However to form the buried tunnel junction (BTJ) an additional growth step is required, making the fabrication of such devices more complicated and expensive.

InGaAs-based VCSELs from NEC have demonstrated excellent reliability [72, 74, 78], confirming the maturity of InGaAs QWs for real world VCSEL applications. Although VCSELs operating at 40 Gbit/s have been demonstrated, these were devices based on the tunnel junction without oxide apertures. The fastest of any oxide-confined VCSELs, which are straightforward in the fabrication and thus inexpensive in the large-scale manufacturing, are currently to the best of our knowledge the 850 nm QW-VCSELs developed by the authors in the framework of this dissertation, which operate at the bit rate of 38 Gbit/s.

1.5 Dissertation Contribution and Overview

The main goal of this dissertation was to investigate and develop high speed directly modulated GaAs-based oxide-confined VCSELs emitting at the wavelengths around 850 and 980 nm for the future ultra-high speed short and ultra-short reach optical interconnects. Special attention was paid to improvement of the thermal stability of the lasers, especially for upcoming applications inside computers. Additionally, straightforward fabrication process compatible with inexpensive scalable mass production technology should be developed, simultaneously enabling high speed reliable operation of the fabricated VCSELs at bit rates of 35 Gbit/s and larger, sufficient enough for the targeted applications. Complicated growth steps should be avoided as well, leaving the overall manufacturing process convenient and inexpensive.

Within the scope of this dissertation three types of VCSELs were investigated: the 980 nm VCSELs based on the InGaAs active layers grown in the submonolayer growth mode (980 nm SML-VCSEL), the 980 nm VCSELs based on the InGaAs QWs (980 nm QW-VCSELs) and the 850 nm VCSELs based on the InGaAs QWs (850 nm QW-VCSELs). While the ultimate goal of the 850 nm QW-VCSELs was to achieve error free data transmission at the largest possible bit rate at room temperature, VCSELs emitting at 980 nm were optimized for the temperature insensitive high speed operation.

The investigation within the framework of this dissertation resulted in the worldwide first 850 nm VCSELs operating error free at the bit rate of 38 Gbit/s. These were also first of any oxide-confined VCSELs operating at such high bit rates. The main achievement of 980 nm SML-VCSELs was the worldwide first demonstration of open eye operation at 20 Gbit/s at elevated temperatures of up to 120°C. These were to the best of our knowledge worldwide first of any VCSELs operating at 20 Gbit/s at such high temperatures. With 980 nm QW-VCSELs open eye operation at 25 Gbit/s at 85°C were achieved, confirming the maturity of the 980 nm VCSELs for the future Datacom application requiring high temperature stability.

The present dissertation consists of six chapters. In the first chapter the overview of data transmission lines and the role of optical technologies, especially of the directly modulated GaAs-based VCSELs, today and expectations for the future are given. Nowadays optics is indispensable for our communication and its role is continuously growing. VCSELs present thereby one of the key basic technologies for successfully development of the future optical interconnects.

In Chap. 2 theoretical background necessary for understanding and successful designing of the high speed laser properties is introduced. Optical, electrical and thermal phenomena are described. An introduction into the rate equation model required for correct analysis of the measured high speed results is given as well. Concepts applied for the development of the high speed VCSELs investigated in this dissertation are presented, together with important simulation results.

Chapter 3 deals with the growth and fabrication techniques applied for the VCSELs developed in this dissertation. Special attention is paid to the description of the straightforward laser fabrication process.

In Chapter 4 results and discussion of the high temperature stable 980 nm VCSELs are presented. First, measurements on the 980 nm SML-VCSELs are described, followed by the results of the 980 nm QW-VCSEL. Continuous wave (CW) and high frequency (HF) measurements are discussed.

Chapter 5 presents measurement results of the high speed 850 nm QW-VCSELs. CW, small and large signal modulation measurements are described. Bit error rate (BER) measurements are presented as well.

Finally, Chap. 6 concludes and summarizes the investigations carried out within the scope of this dissertation and gives an outlook for the future works both on improvements of the VCSELs described in this dissertation and on further developments of high speed devices for future optical interconnects.

Appendix presents description of the measurement setups used for investigations in the scope of this dissertation. Setups for measurement of static and dynamic VCSEL characteristics and used equipment are described in details.

References

1. http://www.bordalierinstitute.com/evolution.html
2. Soda H, Iga K, Kitahara C, Suematsu Y (1979) GaInAsP/InP surface emitting injection lasers. Jpn J Appl Phys 18(12):2329–2330
3. Iga K, Koyama F, Kinoshita S (1988) Surface emitting semiconductor lasers. IEEE J Quant Electron 24(9):1845–1855
4. Lee YH, Tell B, Brown-Goebeler KF, Leibenguth RE, Mattera VD (1991) Deep-red continuous wave top-surface-emitting vertical-cavity AlGaAs superlattice lasers. IEEE Photon Technol Lett 3(2):108–109
5. Iga K (2000) Surface-emitting laser—its birth and generation of new optoelectronics field. IEEE J Sel Top Quant Electron 6(6):1201–1215
6. Iga K (2008) Vertical-cavity surface-emitting laser: its conception and evolution. Jpn J Appl Phys 47(1):1–10
7. Collins D, Li N, Kuchta D, Doany F, Schow C, Helms C, Yang L (2008) Development of high-speed VCSELs: 10 Gb/s serial links and beyond. In: Proceedings of the SPIE 6908-09
8. Doany FE, Schares L, Schow CL, Schuster C, Kuchta DM, Pepeljugoski PK (2006) Chip-to-chip optical interconnects. In: OFC, Anaheim, CA, USA, OFA3
9. Ebeling KJ, Michalzik R, King R, Schnitzer P, Wiedemann D, Jäger R, Jung C, Grabherr M, Miller M (1998) Applications of VCSELs for optical interconnects. In: Proceedings of the 24th European conference on optical communication, Madrid, Spain, vol 3, pp 29–31
10. Ahadian J, Kusumoto K. Analog modulation characteristics of multimode fiber links based on commercial VCSELs. Executive summary, Ultra Communications, Inc. http://ultracomm-inc.com/Documentation/RF-Photonics-Summary.pdf. Accessed 11 August 2009
11. Kash J. Internal optical interconnects in next generation high-performance servers. http://www.cns.cornell.edu/documents/JeffKashIBMTJWatsonResearchCenter.pdf

12. Huffaker DL, Deppe DG, Kumar K, Rogers TJ (1994) Native-oxide defined ring contact for low threshold vertical-cavity lasers. Appl Phys Lett 65:97–99
13. Choquette KD, Schneider RP Jr, Lear KL, Geib KM (1994) Low threshold voltage vertical-cavity lasers fabricated by selective oxidation. Electron Lett 30(24):2043–2044
14. Koyama F (2006) Recent advances of VCSEL photonics. J Lightwave Technol 24(12):4502–4513
15. Avago Technologies. http://www.avagotech.com
16. Finisar Corporation. http://www.finisar.com
17. Emcore Corporation. http://www.emcore.com/
18. Benner A (2009) Cost-effective optics: enabling the exascale roadmaps. In: 17th annual IEEE symposium on high-performance interconnects, New York. http://www.hoti.org/hoti17/program/slides/SpecialSession/Benner_Optics_Enabling_Exascale_Roadmaps.HotI.090827.pdf. Accessed 27 August 2009
19. Emcore AOC product brief. http://www.emcore.com/fiber_optics/emcoreconnects
20. Active optical cables market report 2010. Information Gatekeepers Inc. http://www.igigroup.com/st/pages/aoc.html. Accessed 1 January 2010
21. Light Peak, Intel. http://www.intel.com/go/lightpeak
22. Savage N (2002) Linking with light. IEEE Spectrum vol 39(8) pp 32–36, Cover story, August
23. Mohammed E, Alduino A, Thomas T, Braunisch H, Lu D, Heck J, Liu A, Young I, Barnett B, Vandentop G, Mooney R (2004) Optical interconnects system integration for ultra-short-reach applications. Intel Technol J 8(2):115–128
24. Kobrinsky MJ, Block BA, Zheng J-F, Barnett BC, Mohammed E, Reshotko M, Robertson F, List S, Young I, Cadien K (2004) On-chip optical interconnects. Intel Technol J 8(2): 129–142
25. Miller DAB (1997) Physical reasons for optical interconnection. Intel J Optoelectron 11: 155–168
26. Kern AM (2007) CMOS circuits for VCSEL-based optical IO. Dissertation, Massachusetts Institute of Technology, June
27. Palermo S (2007) Design of high-speed optical interconnect transceivers. Dissertation, Stanford University, September
28. Benner AF, Ignatowski M, Kash JA, Kuchta DM, Ritter MB (2005) Exploitation of optical interconnects in future server architectures. IBM J Res Dev 49(4/5):755
29. Schares L, Kash J, Doany F, Schow CL, Schuster C, Kuchta DM, Pepeljugoski PK, Trewhella JM, Baks CW, John RA, Shan L, Kwark YH, Budd RA, Chiniwalla P, Libsch FR, Rosner J, Tsang CK, Patel CS, Schaub JD, Dangel R, Horst F, Offrein BJ, Kucharski D, Guckenberger D, Hedge S, Nyikal H, Lin C-K, Tandon A, Trott GR, Nystrom M, Bour DP, Tan RTM, Dolfi DW (2006) Terabus: terabit/second-class card-level optical interconnects technologies. IEEE J Sel Top Quant Electron 12(5):1032–1044
30. Kash JA, Doany F, Kuchta D, Pepeljugoski P, Schares L, Schaub J, Schow C, Trewhella J, Baks C, Kwark Y, Schuster C, Shan L, Tsang C, Rosner J, Libsch F, Budd R, Chiniwalla P, Guckenberger D, Kucharski D, Dangel R, Offrein B, Tan M, Troff G, Lin D, Tandon A, Nystrom M (2005) Terabus: a chip-to-chip parallel optical interconnects. In: LEOS 2005, the 18th annual meeting of the IEEE Lasers & Electro-Optics Society, Hilton Sydney, Sydney, Australia, TuW3, 23–27 October 2005, 14:30–14:45
31. Lemoff BE, Ali ME, Panotopoulos G, Flower GM, Madhavan B, Levi AFJ, Dolfi DW (2004) MAUI: enabling fiber-to-the-processor with parallel multi-wavelength optical interconnects. J Lightw Technol 22(9):2043–2054
32. Whitman B (2004) International FTTH deployments: lessons learned around the globe. In: FTTH conference 2004, Orlando, FL, USA, 6 October 2004
33. Pavesi L, Guillot G (eds) (2006) Optical interconnects: the silicon approach. Springer, Berlin
34. Fang AW, Lively E, Kuo Y-H, Liang D, Bowers JE (2008) Distributed feedback silicon evanescent laser. In: Proceedings of the optical fiber communication conference
35. Fang AW (2008) Silicon evanescent lasers. Dissertation, University of California, Santa Barbara, CA, USA, March

36. Fang AW, Lively E, Kuo Y-H, Liang D, Bowers JE (2008) A distributed feedback silicon evanescent laser. Optics Express 16(7):4413–4419
37. Fang AW, Jones R, Park H, Cohen O, Raday O, Paniccia MJ, Bowers JE (2007) Integrated AlGaInAs-silicon evanescent racetrack laser and photodetector. Optics Express 15(5):2315–2322
38. Koch BR, Fang AW, Cohen O, Bowers JE (2007) Mode-locked silicon evanescent lasers. Optics Express 15(18):11225–11233
39. Fang AW, Park H, Cohen O, Jones R, Paniccia MJ, Bowers JE (2006) Electrically pumped hybrid AlGaInAs-silicon evanescent laser. Optics Express 14(20):9203–9210
40. Park H, Fang AW, Cohen O, Jones R, Paniccia MJ, Bowers JE (2007) A hybrid AlGaInAs-silicon evanescent amplifier. IEEE Photonics Technol Lett 19(4):230–232
41. Yin T, Cohen R, Morse M, Sarid G, Chetrit Y, Rubin D, Paniccia M (2008) 40 Gb/s Ge-on-SOI waveguide photodetectors by selective Ge growth. In: Proceedings of the optical fiber communication conference
42. Liao L, Liu A, Rubin D, Basak J, Chetrit Y, Nguyen H, Cohen R, Izhaky N, Paniccia M (2007) 40 Gbit/s silicon optical modulator for high-speed applications. Electron Lett 43(22):1196–1197
43. Moore's Law. http://www.intel.com/technology/mooreslaw
44. Moore GE (1965) Cramming more components onto integrated circuits. The experts look ahead. Electronics 38(8)
45. Press release (2005) Innovation more important than ever in platform era. Intel Developer Forum, San Francisco, CA, USA, 1 March
46. International technology roadmap for semiconductors, 2007 edn (ITRS 2007). Executive Summary http://www.itrs.net/Links/2007ITRS/ExecSum2007.pdf
47. Jewell J, Graham L, Crom M, Maranowski K, Smith J, Fanning T (2006) 1310 nm VCSELs in 1–10 Gb/s commercial application. In: Proceedings of the SPIE, vol 6132, p 613204
48. Hofmann W, Müller M, Nadtochiy AM, Meltzer C, Mutig A, Böhm G, Rosskopf J, Bimberg D, Amann MC, Chang-Hasnain C (2009) 22 Gb/s long wavelength VCSELs. Optics Express 17(20):17547–17554
49. Leisher PO, Chen C, Sulkin JD, Alias MSB, Sharif KAM, Choquette KD (2007) High modulation bandwidth implant-confined photonic crystal vertical-cavity surface-emitting lasers. IEEE Photonics Technol Lett 19(19):1541–1543
50. Danner AJ, Raftery JJ Jr, Leisher PO, Choquette KD (2006) Single mode photonic crystal vertical cavity lasers. Appl Phys Lett 88:091114
51. Leisher PO, Danner AJ, Raftery JJ Jr, Siriani D, Choquette KD (2006) Loss and index guiding in single-mode proton-implanted holey vertical-cavity surface-emitting lasers. IEEE J Quant Electron 42(10):1091–1096
52. Lott JA, Shchukin VA, Ledentsov NN, Stintz A, Hopfer F, Mutig A, Fiol G, Bimberg D, Blokhin SA, Karachinsky LY, Novikov II, Maximov MV, Zakharov ND, Werner P (2009) 20 Gbit/s error free transmission with ~850 nm GaAs-based vertical cavity surface emitting lasers (VCSELs) containing InAs–GaAs submonolayer quantum dot insertions. In: Proceedings of the SPIE 7211, paper 7211-40, Photonics West 2009, San Jose, CA, 29 January 2009
53. Johnson RH, Kuchta DM (2008) 30 Gb/s directly modulated 850 nm Datacom VCSELs. In: Conference on lasers and electro-optics (CLEO), CLEO Postdeadline Session II (CPDB), San Jose, CA, 4 May 2008
54. Westbergh P, Gustavsson JS, Haglund A, Sköld M, Joel A, Larsson A (2009) High-speed, low-current-density 850 nm VCSELs. IEEE J Sel Top Quant Electron 15(3):694–703
55. Westbergh P, Gustavsson JS, Haglund A, Larsson A, Hopfer F, Fiol G, Bimberg D, Joel A (2009) 32 Gbit/s multimode fiber transmission using high-speed, low current density 850 nm VCSEL. IEEE Electron Lett 45(7)
56. Ko J, Hegblom ER, Akulova Y, Thibeault BJ, Coldren LA (1997) Low-threshold 840-nm laterally oxidized vertical-cavity lasers using AlInGaAs–AlGaAs strained active layers. IEEE Photon Technol Lett 9(7):863–865

57. Kuo HC, Chang YS, Lai FY, Hsueh TH, Laih LH, Wang SC (2003) High-speed modulation of 850 nm InGaAsP/InGaP strain-compensated VCSELs. Electron Lett 39(14):1051–1053
58. Kuchta DM, Pepeljugoski P, Kwark Y (2001) VCSEL modulation at 20 Gb/s over 200 m of multimode fiber using a 3.3 V SiGe laser driver IC. Technical digest LEOS summer topical meeting, paper no. WA1.2, pp 49–50
59. Westbergh P, Gustavsson J, Haglund A, Sunnerud H, Larsson A (2008) Large aperture 850 nm VCSELs operating at bit rates up to 25 Gbit/s. Electron Lett 44(15):907–908
60. Westbergh P, Gustavsson JS, Haglund A, Larsson A (2008) Large aperture 850 nm VCSEL operating at 28 Gbit/s. In: The 21st IEEE international semiconductor laser conference 2008, MB1, Sorrento, Italy, September 2008
61. Blokhin SA, Lott JA, Mutig A, Fiol G, Ledentsov NN, Maximov MV, Nadtochiy AM, Shchukin VA, Bimberg D (2009) Oxide-confined 850 nm VCSELs operating at bit rates up to 40 Gbit/s. Electron Lett 45(10)
62. Mutig A, Blokhin SA, Nadtochiy AM, Fiol G, Lott JA, Shchukin VA, Ledentsov NN, Bimberg D (2009) Frequency response of large aperture oxide-confined 850 nm vertical cavity surface emitting lasers. Appl Phys Lett 95:131101
63. Kash JA, Doany FE, Schares L, Schow CL, Schuster C, Kuchta DM, Pepeljugoski PK, Trewhella JM, Baks CW, John RA, Shan L, Kwark YH, Budd RA, Chiniwalla P, Libsch FR, Rosner J, Tsang CK, Patel CS, Schaub JD, Kucharski D, Guckenberger D, Hegde S, Nyikal H, Dangel R, Horst F (2006) Chip-to-chip optical interconnects. In: Optical fiber communication conference and exposition and the national fiber optic engineers conference (OFC), optical interconnect technology (OFA), paper OFA3, Anaheim, CA, 5 March 2006
64. Chang YC, Wang CS, Coldren LA (2007) High-efficiency high-speed VCSELs with 35 Gbit/s error-free operation. IEEE Electron Lett 43(19):1022–1023
65. Chang Y-C (2008) Engineering vertical-cavity surface-emitting lasers for high-speed operation. Thesis, University of California Santa Barbara, December
66. Hopfer F, Mutig A, Fiol G, Kuntz M, Shchukin V, Ledentsov NN, Bimberg D, Mikhrin SS, Krestnikov IL, Livshits DA, Kovsh AR, Bornholdt C (2006) 20 Gb/s 85°C error free operation of VCSEL based on submonolayer deposition of quantum dots. In: IEEE-LEOS 20th international semiconductor laser conference (ISLC), Kohala Coast, HI, USA, 17–21 September 2006
67. Hopfer F, Mutig A, Fiol G, Kuntz M, Shchukin VA, Haisler VA, Warming T, Stock E, Mikhrin SS, Krestnikov IL, Livshits DA, Kovsh AR, Bornholdt C, Lenz A, Eisele H, Dähne M, Ledentsov NN, Bimberg D (2007) 20 Gb/s 85°C error-free operation of VCSELs based on submonolayer deposition of quantum dots. IEEE J Sel Top Quant Electron 13(5)
68. Mutig A, Fiol G, Moser P, Arsenijevic D, Shchukin VA, Ledentsov NN, Mikhrin SS, Krestnikov IL, Livshits DA, Kovsh AR, Hopfer F, Bimberg D (2008) 120°C 20 Gbit/s operation of 980 nm VCSEL. Electron Lett 44(22)
69. Mutig A, Fiol G, Pötschke K, Moser P, Arsenijevic D, Shchukin VA, Ledentsov NN, Mikhrin SS, Krestnikov IL, Livshits DA, Kovsh AR, Hopfer F, Bimberg D (2009) Temperature-dependent small-signal analysis of high-speed high-temperature stable 980-nm VCSELs. IEEE J Sel Top Quant Electron 15(3)
70. Lin C-K, Tandon A, Djordjev K, Corzine SW, Tan MRT (2007) High-speed 985 nm bottom-emitting VCSEL arrays for chip-to-chip parallel optical interconnects. IEEE J Sel Top Quant Electron 13(5):1332–1339
71. Anan T, Suzuki N, Yashiki K, Fukatsu K, Hatakeyama H, Akagawa T, Tokutome K, Tsuji M (2007) High-speed InGaAs VCSELs for optical interconnections. In: International symposium on VCSELs and integrated photonics, Tokyo, Japan, 17–18 December 2007, pp 76–78
72. Suzuki N, Anan T, Hatakeyama H, Fukatsu K, Tokutome K, Akagawa T, Tsuji M (2009) High speed 1.1-μm-range InGaAs-based VCSELs. IEICE Trans Electron E92-C(7):942–950
73. Suzuki N, Hatakeyama H, Fukatsu K, Anan T, Yashiki K, Tsuji M (2006) 25-Gbps operation of 1.1-μm-range InGaAs VCSELs for high-speed optical interconnections. In: Proceedings of the optical fiber communication conference 2006 (OFC), paper no. OFA4

74. Hatakeyama H, Akagawa T, Fukatsu K, Suzuki N, Yashiki K, Tokutome K, Anan T, Tsuji M (2008) 25 Gbit/s-100°C operation and high reliability of 1.1-μm-range VCSELs with InGaAs/GaAsP strain-compensated MQWs. In: Conference on lasers and electro-optics (CLEO), VCSEL I (CMW), San Jose, CA, 4 May 2008
75. Fukatsu K, Shiba K, Suzuki Y, Suzuki N, Hatakeyama H, Anan T, Yashiki K, Tsuji M (2007) 30-Gbps transmission over 100 m-MMFs (GI32) using 1.1 μm-range VCSELs and receivers. In: IEEE 19th international conference on indium phosphide & related materials, IPRM '07, Matsue, Japan, 14–18 May 2007
76. Yashiki K, Suzuki N, Fukatsu K, Anan T, Hatakeyama H, Tsuji M (2007) 1.1-μm-range tunnel junction VCSELs with 27-GHz relaxation oscillation frequency. In: Proceedings of the optical fiber communication conference 2007, paper no. OMK1
77. Yashiki K, Suzuki N, Fukatsu K, Anan T, Hatakeyama H, Tsuji M (2007) 1.1-μm-range low-resistance InGaAs quantum-well vertical-cavity surface-emitting lasers with a buried type-II tunnel junction. Jpn J Appl Phys 46:L512–L514
78. Suzuki N, Hatakeyama H, Tokutome K, Yamada M, Anan T, Tsuji M (2005) 1.1 μm range InGaAs VCSELs for high-speed optical interconnections. In: Proceedings of the lasers and electro-optics society, 2005, paper no. TuAA1, pp 394–395

Chapter 2
Physical Processes in Lasers and VCSEL Design

Semiconductor laser physics is both a very complicated and at the same time a very exciting field. Many excellent books about the physics of semiconductor lasers and other photonic devices have been published, among other [1–4]. Physical processes taking place inside of a laser chip are of intricate nature. Optical, electrical and thermal phenomena have equal importance for the laser operation. Also mechanical phenomena, for example stress, should be considered. Moreover, these phenomena are not independent from each other but build in their strong and commonly non-linear interactions a complex picture of the physical processes inside of a device. For example current flow generates heat, temperature changes affect among other electrical resistance and refractive index of semiconductor materials, presenting photons change carrier concentrations due to absorption and so on. Many physical processes naturally arise from interactions between different fields or particles. For example generation and absorption of the light is an opto-electrical interaction phenomenon between photons and electron–hole pairs. Nonradiative recombination is an interaction process between carriers and phonons, thus an electro-thermal process. For VCSELs the interactions between various physical processes become even stronger than for edge emitting lasers because of their smaller volume, where some dimensions become comparable to the wavelength of the emitting light. For a correct understanding of the VCSEL operation optical, electrical and thermal processes inside of the laser should be considered together, including all important interactions. In Fig. 2.1 interactions between different physical phenomena inside of a VCSEL are shown.

Additionally, because in the modern semiconductor lasers active regions utilizing quantum effects, e.g. QWs or quantum dots (QDs), are implemented, the correct description of the physical processes like photon emission or absorption should be carried out using quantum mechanics. To make the picture even more complicated, time dependence could be introduced, since one is often interested in the modulation properties of lasers.

A. Mutig, *High Speed VCSELs for Optical Interconnects*, Springer Theses,
DOI: 10.1007/978-3-642-16570-2_2, © Springer-Verlag Berlin Heidelberg 2011

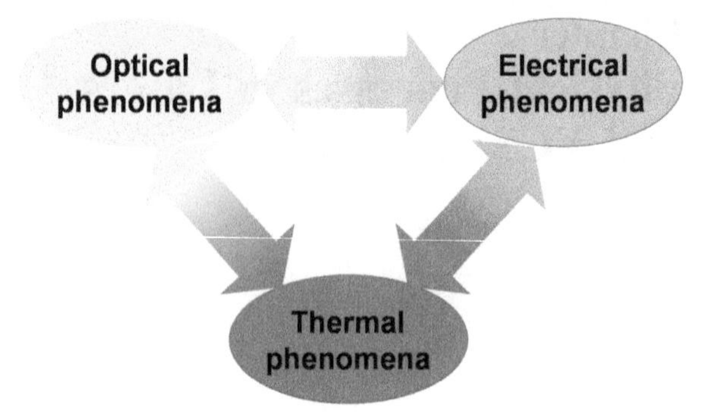

Fig. 2.1 Optical, electrical and thermal phenomena in a semiconductor laser and their interaction

Ideally, a complete opto-electro-thermal three-dimensional model with quantum mechanical description of the active region should be used in order to get the complete picture of the physical processes inside of a VCSEL. In fact such advanced models, dealing with complete or partially coupled opto-electro-thermal phenomena, exist both commercially [5–8] and also in the research groups [9]. Unfortunately their application is very recourse-consuming and usually requires powerful super-computers equipped with parallel processors and enormous amount of memory. In addition, to operate such software efficiently is mostly time-consuming and requires a lot of experience in complicated theoretical simulations of semiconductor devices.

Fortunately, in order to understand basic principles of VCSEL operation, to be able to analyse measurement results efficiently and to design new devices correctly, the application of the complex three-dimensional coupled opto-thermo-electrical models is not implicit necessary. Having a deeper understanding of the physical processes inside of the laser and knowing the basic semiconductor laser theory enable to solve the tasks, that are important for VCSEL designing and analysis, using simpler models. In this case optical, electrical and thermal properties of a VCSEL could be treated separately and all necessary interactions between these phenomena could be considered manually.

In this work optical, electrical and thermal properties of VCSELs have been investigated using decoupled models. Physical parameters, which have been extracted from these individual models, have been then used for device designing and analysis under application of the rate equation model for photon and carrier dynamics.

2.1 Optical Properties

The main purpose of any laser is to generate light, and light is an electro-magnetic wave and thus an optical phenomenon. That is why the optical properties of any laser are playing an important role in the laser physics. In many cases one-dimensional treatment of the optical fields is sufficient. From the other side, many important laser characteristics, especially in the case of VCSELs, could not be calculated from one-dimensional models, making three-dimensional simulations necessary. The presence of the rotation symmetry in VCSELs simplifies the situation, thus effectively two-dimensional optical simulations could be sufficient.

A commonly used model for one-dimensional calculations is the transmission or transfer matrix method [2], which is very fast and easy to use. This model enables easily to calculate among other reflectivity spectra of the multilayered distributed Bragg reflectors (DBR), which act as top and bottom mirrors, longitudinal standing wave pattern inside of the VCSEL cavity and the cavity dip position. This basic information is inalienable for the correct VCSEL designing. Nevertheless for a deeper understanding of the optical laser properties, especially for oxide-confined VCSELs, three-dimensional field distributions for the ground and also higher order lasing modes as well as important mode parameters, e.g. mode volume and optical confinement factor, should be accessible. Such kind of calculations could be carried out only by three-dimensional or, in the case of the presence of rotation symmetry, also by two-dimensional models.

Several models for two-dimensional and three-dimensional simulations exist and can be divided into scalar, vectorial and hybrid models. Scalar models solve scalar Helmholtz equation and are commonly fast, but not very accurate in handling the higher order lasing modes or devices with smaller dimensions. Vectorial models solve Maxwell's equations exactly, without any approximations. These models are slower than scalar methods but can handle every optical field, including also those with a relatively large transverse wavevector component, like in the case of higher order modes or smaller devices, accurately. Hybrid models are in between and apply both scalar and vectorial approaches to deal with different optical subproblems.

Commonly used scalar models are the effective index model [10], the effective frequency method [11] and the effective index model with eigenmodes [12]. From the vectorial models the Green's function model [13], the full-vector weighted index method [14] and the eigenmode expansion with perfectly matched layers (PML) [15, 16], are commonly used. An overview and a comparison of different optical models could be found in [17].

In this work both the one-dimensional transfer matrix method and the three-dimensional eigenmode expansion technique with PMLs, both implemented in the free CAMFR software [18] were applied.

2.1.1 Transfer Matrix Method and 1D Simulations

The general equations describing electro-magnetic fields are the Maxwell's equations [1], which relate electric field $\vec{E}$, magnetic field $\vec{H}$, electric displacement flux density $\vec{D}$ and magnetic flux density $\vec{B}$ to the charge density ρ and the current density $\vec{J}$ (2.1.1–2.1.4).

$$\vec{\nabla} \times \vec{E} = -\frac{\partial}{\partial t}\vec{B} \tag{2.1.1}$$

$$\vec{\nabla} \times \vec{H} = \vec{J} + \frac{\partial}{\partial t}\vec{D} \tag{2.1.2}$$

$$\vec{\nabla} \cdot \vec{D} = \rho \tag{2.1.3}$$

$$\vec{\nabla} \cdot \vec{B} = 0. \tag{2.1.4}$$

Thereby the two source terms, the charge density and the current density, are not independent but related by the continuity equation (2.1.5).

$$\vec{\nabla} \cdot \vec{J} + \frac{\partial}{\partial t}\rho = 0 \tag{2.1.5}$$

The quantities $\vec{E}$, $\vec{H}$, $\vec{D}$ and $\vec{D}$ are also not independent but related by the constitutive relations, that involve the properties of the medium:

$$\vec{D} = \varepsilon \cdot \vec{E}, \tag{2.1.6}$$

$$\vec{B} = \mu \cdot \vec{H}, \tag{2.1.7}$$

where ε is in a common case the permittivity tensor and μ is the permeability tensor. For an isotropic media ε and μ are scalars.

For time-harmonic fields it is common to apply the so called phasor notation, where the harmonic time dependence is represented by the complex exponents of the form $e^{j\omega t}$ and j is the imaginary unit. The real time-dependent fields can be then extracted from the complex phasors using Eqs. 2.1.8 and 2.1.9:

$$\vec{E}(\vec{r}, t) = \mathrm{Re}\left\{\vec{E}(\vec{r}, \omega) \cdot e^{j\omega t}\right\}, \tag{2.1.8}$$

$$\vec{H}(\vec{r}, t) = \mathrm{Re}\left\{\vec{H}(\vec{r}, \omega) \cdot e^{j\omega t}\right\}. \tag{2.1.9}$$

Here $\vec{E}(\vec{r}, t)$ and $\vec{H}(\vec{r}, t)$ are the real electric and magnetic fields and $\vec{E}(\vec{r}, \omega)$ and $\vec{H}(\vec{r}, \omega)$ the complex phasors, which are not time-dependent. In the further equations we will use the symbols $\vec{E}$, $\vec{H}$ etc. for phasors, if nothing special is noted. All other quantities could be easily converted to the phasors and back using the same rules.

The phasor notation simplifies mathematical calculations and representation of simulation results. Maxwell equations for a homogeneous isotropic medium without any charges and current flow ($\vec{J} = 0$ and $\rho = 0$) could be then written down using phasors and substituting $\vec{D}$ and $\vec{B}$ from Eqs. 2.1.6 and 2.1.7 as follows:

$$\vec{\nabla} \times \vec{E} = -j\omega\mu\vec{H}, \tag{2.1.10}$$

$$\vec{\nabla} \times \vec{H} = \vec{J} + j\omega\varepsilon\vec{E}, \tag{2.1.11}$$

$$\vec{\nabla} \cdot \vec{D} = \rho, \tag{2.1.12}$$

$$\vec{\nabla} \cdot \vec{B} = 0. \tag{2.1.13}$$

Equations 2.1.6 and 2.1.7 stay unchanged.

The solutions of the Maxwell's equation in an isotropic dielectric medium are plane waves of the form

$$\vec{E}_f = \vec{e}E_f e^{-\beta z}, \tag{2.1.14}$$

$$\vec{E}_b = \vec{e}E_b e^{\beta z}, \tag{2.1.15}$$

for the forward propagating wave $\vec{E}_f$ and the backwards propagating wave $\vec{E}_b$. The unit vector $\vec{e}$ determines the polarization of the light. The propagation direction is parallel to the z-axis. The quantity β is the propagation constant and for the plane wave in an isotropic dielectric medium is determined by the refractive index of the medium n:

$$\beta = \omega n/c. \tag{2.1.16}$$

Transfer matrix formalism handles plane waves and use the fact that each interface between two isotropic dielectric media as well as each layer could be represented by 2×2 matrices, which are called transmission or transfer matrices T. Each transmission matrix relates forward and backward propagating field amplitudes $E_{1,f}$ and $E_{1,b}$ on one side of the interface or layer with the forward and backward propagating fields $E_{2,f}$ and $E_{2,b}$ on the other side:

$$\begin{pmatrix} E_{1,f} \\ E_{1,b} \end{pmatrix} = \begin{pmatrix} T_{11} & T_{12} \\ T_{21} & T_{22} \end{pmatrix} \begin{pmatrix} E_{2,f} \\ E_{2,b} \end{pmatrix} = T \begin{pmatrix} E_{2,f} \\ E_{2,b} \end{pmatrix}. \tag{2.1.17}$$

The great advantage comes into play by using the transfer matrix method for multilayered structures. For the complete stack of layers one can simple multiply the corresponding T_i matrices of each individual layer and interface and so get the T_S matrix for the whole stack:

$$T_S = T_1 \cdot T_2 \cdot \cdots \cdot T_i \cdot \cdots \cdot T_{N-1} \cdot T_N, \tag{2.1.18}$$

where N is the total number of the layers and interfaces. This idea together with the definitions of the forward and backward directions is illustrated in Fig. 2.2.

The transmission matrices can be easily calculated for plane waves and for the case of the normal incidence for an interface between two media with refractive indices $n_1$and n_2, and also for a layer with the thickness L according to Eqs. 2.1.19 and 2.1.20, respectively,

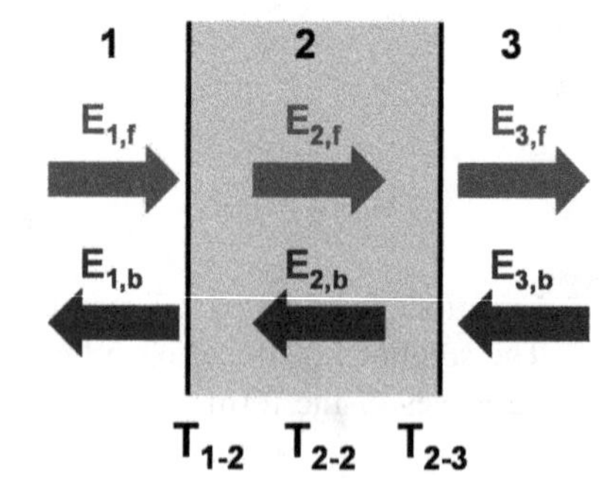

Fig. 2.2 Transmission matrices for a stack consisting of three media 1, 2 and 3 and two interfaces

$$T_{\text{interface}} = \frac{1}{t_{12}} \begin{pmatrix} 1 & r_{12} \\ r_{12} & 1 \end{pmatrix}, \tag{2.1.19}$$

$$T_{\text{layer}} = \begin{pmatrix} e^{-j\beta L} & 0 \\ 0 & e^{j\beta L} \end{pmatrix}, \tag{2.1.20}$$

with r_{12} and t_{12} given by the following equations:

$$r_{12} = \frac{n_1 - n_2}{n_1 + n_2}, \tag{2.1.21}$$

$$t_{12} = \frac{2n_1}{n_1 + n_2}. \tag{2.1.22}$$

Transfer matrix method is very simple to use and do not require noticeable computational power. Using this formalism in the present work 1D-simulations for DBR reflectivity spectra, cavity dip position and field distribution inside of the cavity were carried out. In the following a short outline of each of these three tasks with exemplary simulation results used for VCSEL designing will be given.

One of the first tasks while designing a VCSEL is to decide, how many DBR pairs for the top and the bottom mirror should be grown. This determines the reflectivity of the both mirrors. Figure 2.3a shows the reflectivity and phase spectra for the top $Al_{0.12}Ga_{0.88}As/Al_{0.90}Ga_{0.10}As$ DBR with 23.5 pairs used in the 980 nm QW-VCSELs. The simulation includes two 30 nm thick $Al_{0.98}Ga_{0.02}As$ layers for the later wet oxidation, in the first DBR pair closest to the cavity and just below it. To minimize electrical resistance 20 nm thick linear gradings were applied and also token into consideration for the simulations.

From Fig. 2.3a can be obtained that the maximum reflectivity and also the zero phase appear at the desired wavelength of 980 nm. Figure 2.3b demonstrates dependence of the peak reflectivity on the number of the DBR pairs. Adding additional two pairs to 21.5 pairs increase the reflectivity at 980 nm from 0.9960 up to 0.9977. For VCSELs reflectivity of the mirrors should be designed with a

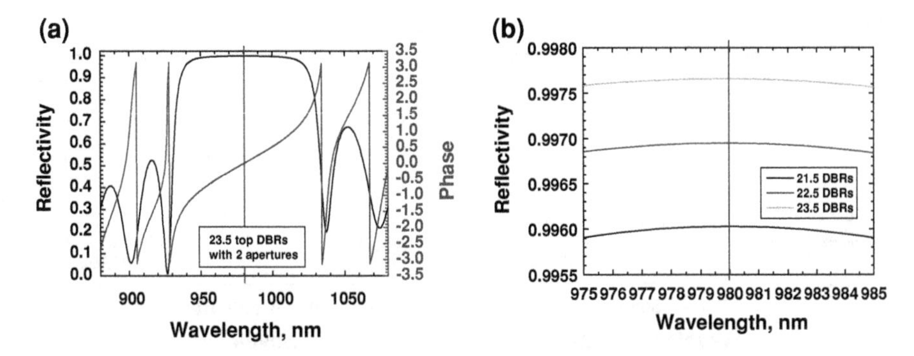

Fig. 2.3 Reflectivity and phase as a function of the wavelength of an $Al_{0.12}Ga_{0.88}As$/$Al_{0.90}Ga_{0.10}As$ DBR with 23.5 DBR pairs including two $Al_{0.98}Ga_{0.02}As$ layers for the later oxidation (**a**) and reflectivity as a function of the number of the DBR pairs (**b**)

very high precision, so that four numbers after the decimal point should be considered. According to the well known equation [1, 2]

$$\Gamma g_{\mathrm{thr}} = \alpha_i + \frac{1}{2L}\ln\left(\frac{1}{R_1 R_2}\right) = \alpha_i + \alpha_m = \alpha_{\mathrm{tot}}, \qquad (2.1.23)$$

where Γ is the optical confinement factor, g_{th} the threshold gain, α_i the intrinsic cavity loss, α_m the mirror loss, α_{tot} the total loss, R_1 and R_2 the power reflection coefficients of the top and bottom mirrors and L the effective cavity length, the reflectivity of each mirror directly affects the threshold gain and thus the threshold current. Reflectivities in the range of 0.997–0.998 for the top mirror and close to 0.9998–0.9999 for the bottom mirror were used for all structures investigated in this work. For the 980 SML-VCSELs 20.5 periods of the GaAs/$Al_{0.90}$GaAs DBRs for the top and 32.5 periods for the bottom mirror with 10 nm thick linear gradings were used. For the 980 nm QW-VCSELs 23.5 and 37.5 $Al_{0.12}Ga_{0.88}As$/$Al_{0.90}Ga_{0.10}As$ DBR pairs for the top and bottom mirrors respectively were grown. For the 850 nm QW-VCSELs these were 22.5 and 35 $Al_{0.15}Ga_{0.85}As$/$Al_{0.90}Ga_{0.10}$ DBR pairs respectively. In the last two structures the thickness of the linear gradings was 20 nm.

The second important task for a proper VCSEL design is to match the cavity wavelength to the desired value. For the VCSEL fabricated in this work these were 850 and 980 nm. For this purpose the position of the cavity dip was calculated with the transfer matrix method. Therefore reflection from the complete VCSEL structure was simulated. If necessary, the thickness of the cavity should be adjusted in order to match the dip position to the desired wavelength. Figure 2.4 shows the calculated cavity dip positions for the 850 nm QW-VCSELs and 980 nm QW-VCSELs. Both dips match the desired wavelengths sufficiently precise.

Finally, the active medium should be placed on the right place in the field intensity anti-node in order to increase the interaction of the active material with

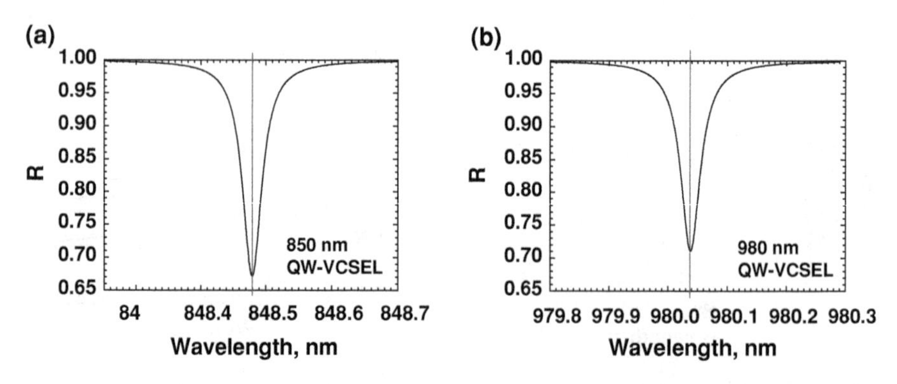

Fig. 2.4 Cavity dip position for the 850 nm QW-VCSEL (**a**) and for the 980 nm QW-VCSEL (**b**)

the optical field. For this purpose standing wave pattern of the ground mode was calculated and investigated. Calculated electrical field magnitudes together with the refractive index profiles inside of the cavity for the 850 nm QW-VCSEL, 980 nm QW-VCSEL and 980 nm SML-VCSEL are shown in Fig. 2.5a–c respectively. The growth direction on the pictures is from the right to the left and at the zero position is the air-semiconductor interface. A zoomed view of the cavity for the 850 nm QW-VCSEL and 980 nm QW-VCSEL is shown in the Fig. 2.6a, b, respectively. From both figures can be obtained that the number of the active layers is odd and the middle active layer is located exactly in the anti-node of the field intensity. The remaining active layers are placed symmetrically to the middle active layer with the possible smallest distance in order to optimize the overlap with the optical field.

As one can see, the transfer matrix method is a simple but at the same time powerful tool to solve basic important tasks for the optical VCSEL designing. Its application is very time-saving and does not require many resources. The simulation time for one calculation is commonly only few seconds, so that many calculations could be carried out within a very short time slot.

In spite of all advantages of the transfer matrix formalism, this method remains limited to one-dimensional problems and plane waves and cannot give a deeper understanding of the optical processes inside of a complex three-dimensional structure like oxide-confined VCSELs, where some dimensions become comparable to the wavelength of the emitting light. Using transfer matrix no three-dimensional field distributions and no corresponding optical parameters, e.g. three-dimensional confinement factor, mode volume etc., can be calculated. There is no direct possibility to handle higher order lasing modes. However, understanding of these phenomena and having access to the three-dimensional properties not only of the ground mode, but also of the higher order modes, is an indispensable prerequisite for a good understanding of the physical processes inside of the laser. For this purpose models, which are able to handle three-dimensional problems, should be applied. In this work the fully-vectorial eigenmode expansion technique handling two-dimensional structures was used, which

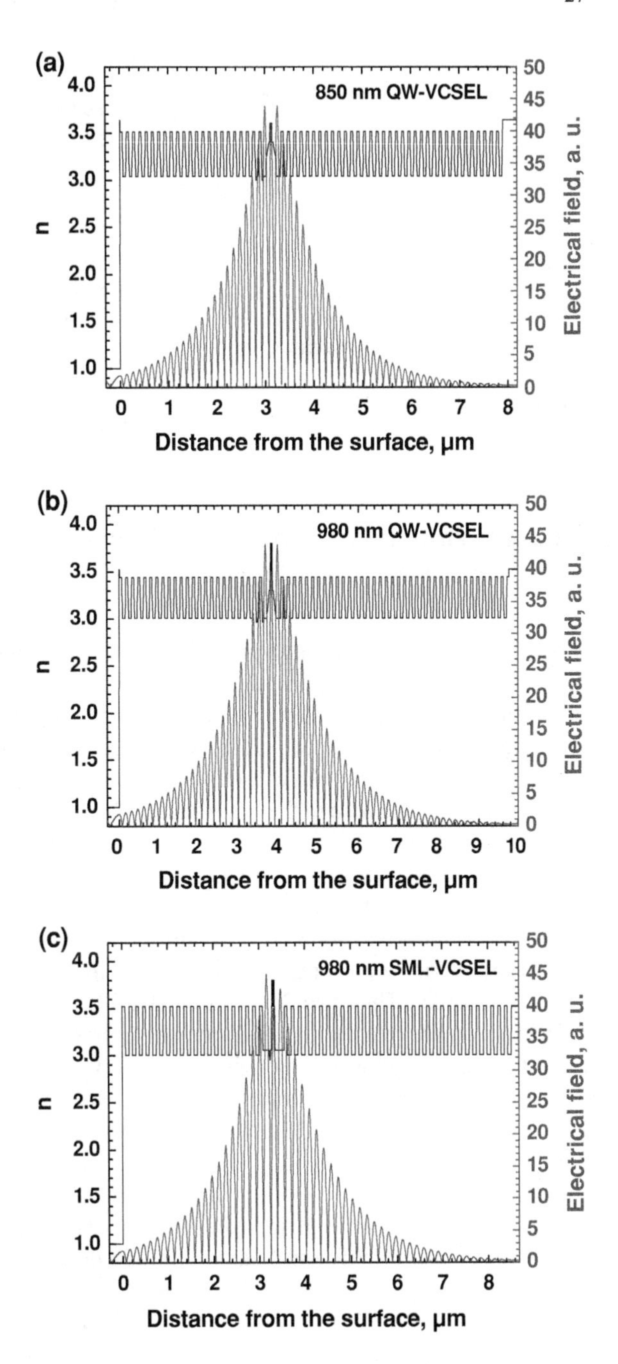

Fig. 2.5 Refractive index profile and electrical field distribution inside of the cavity for the 850 nm QW-VCSEL (**a**), 980 nm QW-VCSEL (**b**) and 980 nm SML-VCSEL (**c**). At zero position is the air–semiconductor interface

could deliver three-dimensional results because of the cylindrical symmetry of the VCSEL structures. This model and the obtained simulation results will be presented in the following sections of this chapter.

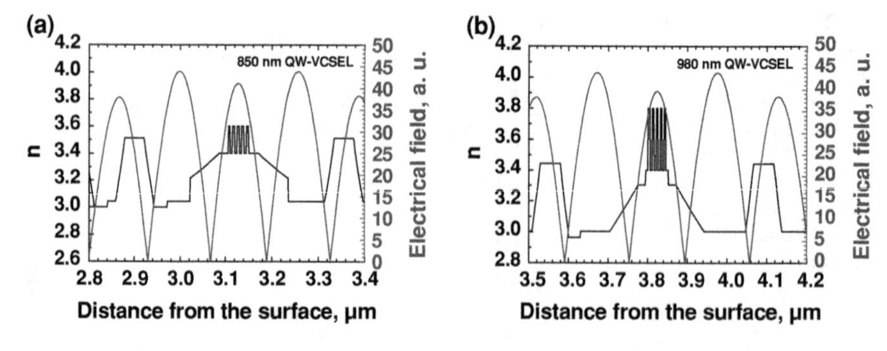

Fig. 2.6 Zoomed view of the refractive index profile and the electrical field distribution inside of the cavity for the 850 nm QW-VCSEL (**a**) and 980 QW-VCSEL (**b**)

2.1.2 Eigenmode Expansion Technique for 3D VCSEL Modeling

Most exact models apply some form of spatial discretisation in order to handle three-dimensional problems. For example for models based on finite elements [19] and finite-difference time-domain (FDTD) methods [20] some kind of grid should be generated and overlays the structure to be modeled. Then in each cell the Maxwell's equations should be solved and the found solutions should be matched to each other at each interface and also to the boundaries. This form of the spatial discretisation is very time- and resource-consuming, since large number of unknowns is generated. Simulations based on such methods require powerful computers and could take many hours or even days.

However, most real VCSEL structures do not have an arbitrary refractive index profile and in fact consist on many layers, but in each of these layers the refractive index is constant. Thus one does not need a full discretisation, and dividing of the VCSEL structure into regions with constant refractive indices, together with taking advantage of the cylindrical symmetry of the VCSEL, is sufficient for a fully three-dimensional optical simulation. Eigenmode expansion technique does exact this. The principle of the spatial discretisation used in the eigenmode expansion method compared to a grid commonly used in other models is schematically shown in Fig. 2.7.

In a z-invariant medium in cylindrical coordinates (Fig. 2.8) one can set the dependence of the electric and magnetic field on the z-coordinate to be harmonic [16]:

$$\vec{E}(\vec{r}) = \left(E_r(r,\phi)\cdot\vec{e}_r + E_\phi(r,\phi)\cdot\vec{e}_\phi + E_z(r,\phi)\cdot\vec{e}_z\right)\cdot e^{-j\beta z}, \tag{2.1.24}$$

$$\vec{H}(\vec{r}) = \left(H_r(r,\phi)\cdot\vec{e}_r + H_\phi(r,\phi)\cdot\vec{e}_\phi + H_z(r,\phi)\cdot\vec{e}_z\right)\cdot e^{-j\beta z}, \tag{2.1.25}$$

with reduced field components E_r, E_ϕ, E_z for the electric and H_r, H_ϕ, H_z for the magnetic field, which depend only on two coordinates: r and ϕ.

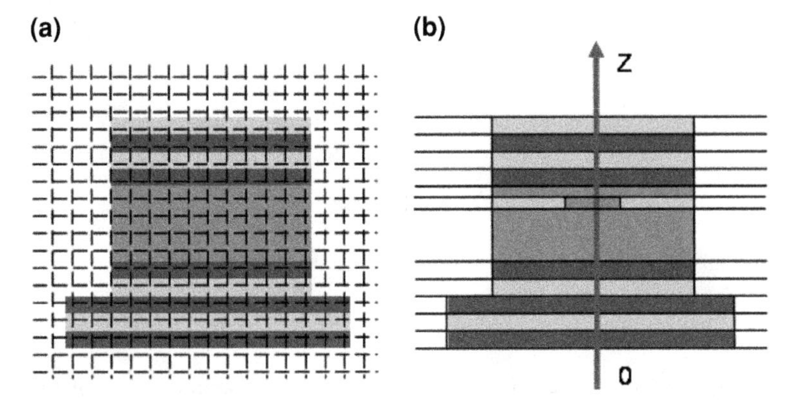

Fig. 2.7 Spatial discretisation with a grid (**a**) and in the eigenmode expansion method (**b**)

Fig. 2.8 A z-invariant medium and cylindrical coordinate system

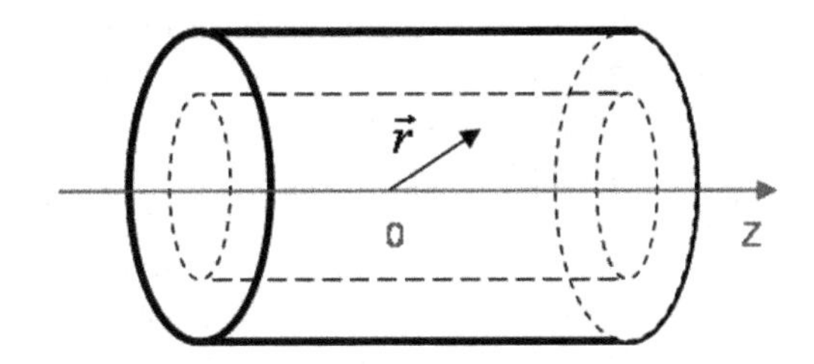

Here β is again the propagation constant and is at the beginning not fixed. With the ansatz (2.1.24) and (2.1.25) following equations for the reduced field components could be derived from the Maxwell's equations (2.1.10–2.1.13) in cylindrical coordinates in absence of electrical charges and currents:

$$\frac{\partial^2 E_z}{\partial r^2} + \frac{1}{r}\frac{\partial E_z}{\partial r} + \frac{1}{r^2}\frac{\partial^2 E_z}{\partial \phi^2} + q^2 E_z = 0, \tag{2.1.26}$$

$$\frac{\partial^2 H_z}{\partial r^2} + \frac{1}{r}\frac{\partial H_z}{\partial r} + \frac{1}{r^2}\frac{\partial^2 H_z}{\partial \phi^2} + q^2 H_z = 0 \tag{2.1.27}$$

for the z-components of the electric and magnetic field and

$$E_r = -\frac{j}{q^2}\left(\beta\frac{z}{\partial r} + \frac{\omega\mu}{r}\frac{\partial H_z}{\partial \phi}\right), \tag{2.1.28}$$

$$E_\phi = -\frac{j}{q^2}\left(\frac{\beta}{r}\frac{\partial E_z}{\partial \phi} - \omega\mu\frac{\partial H_z}{\partial r}\right), \tag{2.1.29}$$

$$H_r = -\frac{j}{q^2}\left(\beta\frac{z}{\partial r} - \frac{\omega\varepsilon}{r}\frac{\partial E_z}{\partial \phi}\right), \tag{2.1.30}$$

$$H_{\phi} = -\frac{j}{q^2}\left(\frac{\beta}{r}\frac{\partial H_z}{\partial \phi} + \omega\varepsilon\frac{\partial E_z}{\partial r}\right) \tag{2.1.31}$$

for the r- and ϕ-components. Hereby is q defined as follows:

$$q = \sqrt{k^2 - \beta^2}, \tag{2.1.32}$$

with k equal to

$$k = \omega\sqrt{\mu\varepsilon}. \tag{2.1.33}$$

One can see from Eqs. 2.1.26 and 2.1.27 that the z-components of the field are decoupled from the r- and ϕ-components. Once one have solved for the z-components one can immediately solve Eqs. 2.1.28–2.1.31 for the r- and ϕ-components and thus get complete fields.

One can make several assumptions regarding the electric and magnetic fields in order to simplify the mathematical procedures and thus get different types of solutions. If one assumes that the z-components of both electric and magnetic fields are equal to zero, the so called transverse electric (TE) and transverse magnetic (TM) modes will be the solutions of (2.1.26–2.1.31). In a common case, where all components of both fields are assumed to be present, the so called hybrid modes (HE and EH) are the desired solutions. For an overview about mode classification in circular cylindrical structures one can refer to [21–23].

For the cylindrical symmetry the ϕ-dependence of the z-components of the electric and magnetic field can be assumed to be as follows:

$$E_z(r,\phi) = A \cdot F_{\nu}(r) \cdot e^{j\nu\phi}, \tag{2.1.34}$$

$$H_z(r,\phi) = B \cdot F_{\nu}(r) \cdot e^{j\nu\phi}, \tag{2.1.35}$$

where A and B are constants, ν is a whole number (0, 1, 2, …) and the function $F_{\nu}(r)$ depends only on the r coordinate and also parametrically on ν. With this ansatz one gets the equation for the function $F_{\nu}(r)$ from (2.1.26) and (2.1.27):

$$\frac{\partial^2 F_{\nu}}{\partial r^2} + \frac{1}{r}\frac{\partial F_{\nu}}{\partial r} + \left(q^2 - \frac{\nu^2}{r^2}\right)F_{\nu} = 0. \tag{2.1.36}$$

The differential equation (2.1.36) is known as the Bessel differential equation and has two linear independent solutions, which are the Bessel function of the first kind $J_{\nu}(qr)$ and the Bessel function of the second kind $Y_{\nu}(qr)$, sometimes called also Weber or Neumann function and noted as $N_{\nu}(qr)$. Also solutions of the modified differential Bessel equation

$$\frac{\partial^2 F_{\nu}}{\partial r^2} + \frac{1}{r}\frac{\partial F_{\nu}}{\partial r} + \left(-w^2 - \frac{\nu^2}{r^2}\right)F_{\nu} = 0 \tag{2.1.37}$$

are important, which are the modified Bessel function of the first kind $I_\nu(wr)$ and the modified Bessel function of the second kind $K_\nu(wr)$, sometimes also known as MacDonald's or Basset function. The number ν in all these solutions represents the order of the function.

Electric and magnetic fields of a z-invariant medium with rotational symmetry could be therefore expressed by the Bessel functions or their linear combinations. Important functions are the so called Henkel functions of the first kind $H_\nu^{(1)}(qr)$ and of the second kind $H_\nu^{(2)}(qr)$, which are defined as follow:

$$H_\nu^{(1)}(qr) = J_\nu(qr) + iY_\nu(qr) \tag{2.1.38}$$

$$H_\nu^{(2)}(qr) = J_\nu(qr) - iY_\nu(qr) \tag{2.1.39}$$

In each region with a constant refractive index, that means according to (2.1.33) with a constant k and thus, according to (2.1.32), also with a constant q, own solutions of the Maxwell's equations of the form (2.1.34) and (2.1.35) are existing. The order of the used functions ν should be thereby in all regions the same because of the boundary conditions on the interfaces between these regions. In the central part of the structure the solutions should be constructed using only the Bessel functions of the first kind, because they are finite at $r = 0$. For the outer regions the modified Bessel functions of the second kind $K_\nu(wr)$ are suitable, because they disappear as r increases: $K_\nu(wr) \underset{wr\to\infty}{\longrightarrow} 0$.

So far we have seen which form electric and magnetic fields in a z-invariant rotational symmetric medium should have. No restrictions were met regarding the propagation constant β, so that for each frequency ω infinite number of solutions was possible. By introducing interfaces in the r direction and also by setting the boundary conditions in this direction the propagation constant β becomes discreet. For a given ν discreet propagation constants exist:

$$\beta_{\nu 1}, \beta_{\nu 2}, \beta_{\nu 3}, \ldots \tag{2.1.40}$$

For each $\beta_{\nu m}$ one solution calling an eigenmode exist. These solutions are denoted by TE_{0m}, TM_{0m}, $HE_{\nu m}$ or $EH_{\nu m}$ respectively. For $\nu = 0$ only TE and TM eigenmodes exist, for $\nu > 0$ only HE and EH modes. The well known linear polarized LP modes are simplifications of the more common case presented here. They could be used if the differences in refractive indices inside of the investigated structure are small. This is mostly not the case for oxide-confined VCSELs, which is why hybrid modes should be used.

After calculation of eigenmodes in each region with constant refractive index, the electric and magnetic fields can be represented as an infinite sum over the forward and backward propagating eigenmodes:

$$\vec{E}(r,\phi,z) = \sum_k \left(A_k^+ \cdot \vec{E}_k(r,\phi) \cdot e^{-j\beta_k z} + A_k^- \cdot \vec{E}_k(r,\phi) \cdot e^{j\beta_k z}\right), \tag{2.1.41}$$

$$\vec{H}(r,\phi,z) = \sum_k \left(A_k^+ \cdot \vec{H}_k(r,\phi) \cdot e^{-j\beta_k z} + A_k^- \cdot \vec{H}_k(r,\phi) \cdot e^{j\beta_k z}\right). \qquad (2.1.42)$$

Here A_k^+ and A_k^- represent expansion coefficients for the forward and backward propagation eigenmode, $\vec{E}_k(r,\phi)$ and $\vec{H}_k(r,\phi)$ are the eigenmode field profiles and β_k is the propagation constant of the eigenmode with index k. In a practical case the number of the used eigenmodes should not be infinite, but a relatively low number of the eigenmodes, depending on the structure and desired precision, is enough to get reasonable results.

Knowing the eigenmodes in every region one can use only the expansion coefficients A_k^+ and A_k^- for the field representation according to Eqs. 2.1.41 and 2.1.42. Thereby in every region own eigenmodes exist and therefore also the expansion coefficients would be different. For interfaces in the z-direction corresponding matrices could be defined, which describe reflection and transmission at each interface, and thus changes of the vector with expansion coefficients while going from one region to another (Fig. 2.9a). For locating the laser modes and calculating of the mode properties, e.g. field distribution, modal volume etc., the VCSEL is divided into two stacks and reflection matrices from both stacks are calculated, as shown in Fig. 2.9b.

A lasing mode should reproduce itself after these two reflections, so that by varying the frequency ω (or the laser light wavelength λ) and the material gain g such field configuration should be found, which satisfies the lasing condition

$$Q = R_{\text{top}} \cdot R_{\text{bot}} = 1. \qquad (2.1.43)$$

The eigenmode expansion technique is an exact and three-dimensional tool, which enables to calculate lasing modes of a "cold" cavity, which means without considering electrical or thermal effects. This method is a pure optical but at the same time a very powerful tool, which gives access to the crucial optical parameters of the ground and also higher order lasing modes. In the next sections of this chapter simulation results for different optical phenomena in VCSELs, obtained with the eigenmode expansion technique realized in the CAMFR freeware, will be presented. All simulations were carried out with the number of modes between 200 and 300. The time for one calculation varied depending on complexity of the desired VCSEL structures, precisions, number of calculated

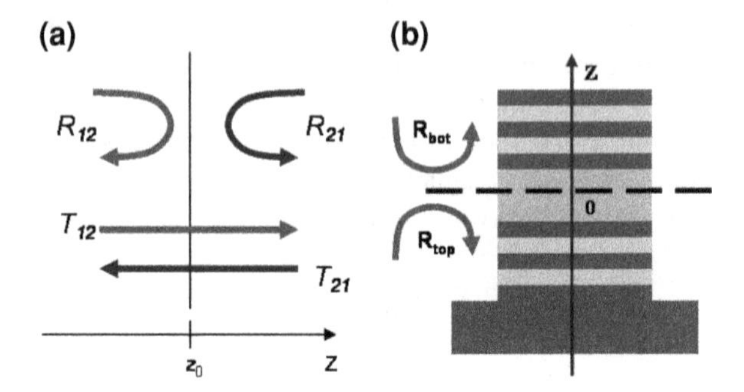

Fig. 2.9 Matrices describing expansion coefficient transformation at interfaces (**a**) and the principle of the laser mode calculation with eigenmode expansion technique (**b**)

pictures etc. between one and several hours, which is a reasonable time effort for a scientist interesting in understanding of important laser phenomena and for efficient VCSEL designing.

2.1.3 Mode Structure of Oxide-Confined VCSELs

Knowing the mode structure and the properties of the ground and higher order lasing modes is of a great importance for an efficient VCSEL designing. Optical properties of laser modes like optical confinement factor and mode volume directly affect CW and modulation characteristics of the fabricated VCSELs. Having a deep understanding of the field distribution for the ground and higher order modes enables to design lasers with proper modal characteristics by using special techniques, for example photonic crystals etched into the top mirror of a VCSEL [24–28], holey VCSELs [29] or VCSELs with surface relief [30–32].

Commonly, VCSELs operate in a single longitudinal mode, but in several transversal modes. In oxide-confined VCSELs, like investigated in this work, optical guiding is realized by the oxide aperture. Oxide-confined VCSELs have established their self because of the reliable and simple fabrication of the oxide aperture and are today commercially available. Mode behavior in oxide-confined VCSELs is strongly dependent on the geometrical parameters of the oxide aperture, e.g. aperture position, thickness, form, diameter and also number of apertures. Using eigenmode expansion technique these effects can be easily calculated for the cold cavity case [33]. Several assumptions should be made in order to be able to use eigenmode expansions. The most important assumption is the homogeneously pumped active region, which is in the reality not the case. The fact that the eigenmode expansion handles only the optical part of a VCSEL restricts the use of the model to achieve quantitative results for physical quantities, which have strongly dependence on electrical and thermal inhomogeneities. Nevertheless for a given temperature and current flow many optical properties of a laser mode, e.g. wavelength, field distribution, mode volume, optical confinement factor etc., remain stable, so that a qualitative picture of the modal behavior of oxide-confined VCSELs could be constructed using fully vectorial three-dimensional optical simulations with the eigenmode expansion technique. This gives a deeper understanding of the physical processes inside of the lasers, which delivers an enormous important contribution to a proper VCSEL design.

First, an overview over the lasing modes existing in an oxide-confined VCSEL will be given. As we have seen in the previous section, in cylindrical structures eigenmodes denoted as TE, TM, HE and EH exist, which have field distributions constructed from the Bessel functions. A laser mode can be expressed in each layer by a sum over the corresponding eigenmodes of this layer. Simulations show, that field profiles of lasing modes have similarities with the eigenmodes. The reason is that in the eigenmode expansion of a lasing mode (2.1.41) and (2.1.42) certain eigenmodes have larger expansion coefficients A_k^+ and A_k^-, and thus play a larger role. Consequently for the classification of the lasing modes similar notation could be used as for circular cylindrical optical fibers [23]. Therefore we denote the lasing modes by HE_{11}, TE_{01} and so on. The first number in the subscript defines the angular dependence of the electric and magnetic field components according to (2.1.34) and (2.1.35), and thus also the order of the Bessel functions used in the eigenmode expansion of this lasing mode. The second number in the subscript denotes the sequence number of the mode. For example for the angular number 1 following lasing modes will exist: HE_{11}, EH_{11}, HE_{12}, EH_{12}, HE_{13}, EH_{13} and so on. We note that for the angular number 0 only TE and TM modes exist, while for angular numbers larger than 0 only HE and EH modes exist. To investigate the mode structure of oxide-confined VCSELs calculations based on a simple VCSEL with one 18.8 nm (1/20 part of the lasing wavelength in GaAs) thick oxide aperture placed in the electric field intensity anti-node, a $3/2\lambda$ thick AlGaAs-based cavity and 25/30 $Al_{0.90}GaAs/GaAs$ top/bottom DBR pairs without any gradings were carried out. The diameter of the oxide aperture was for these simulations 8 μm, enabling lasing of several higher order modes. The diameter of the homogonously pumped active region was also 8 μm. Since the calculations were made for the cold cavity case, the absolute numbers for the threshold material gain should be handled with care. The ground mode and several higher order modes, altogether 21 modes, for the angular numbers from 0 up to 7 were calculated. Obtained wavelength and threshold material gain for each calculated mode are shown in Fig. 2.10.

From the picture is seen, that the lasing mode with the longest wavelength and with the smallest threshold material gain is the HE_{11} mode, which is the ground mode of the laser. Then three modes (TE_{01}, TM_{01} and HE_{21}) with practically identical wavelengths (1093.63, 1093.65 and 1093.64 nm, respectively) follow. In scalar models these three modes are not resolved and handled as one LP_{11} mode. As we will see later these three modes have very similar radial intensity distribution, but different field profiles. Then other lasing modes

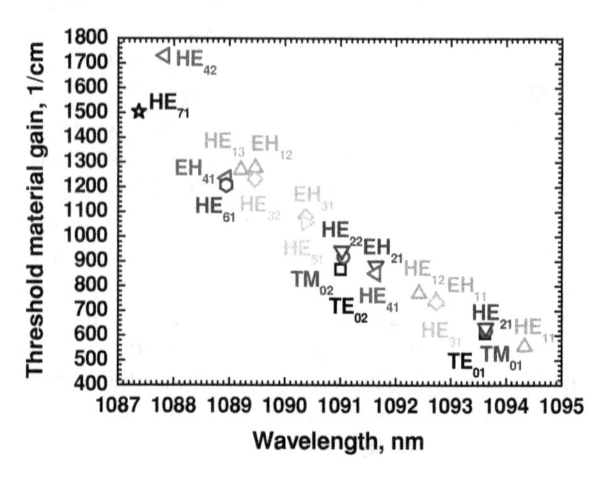

Fig. 2.10 Calculated wavelength and threshold material gain for a number of lasing modes for an oxide-confined VCSEL emitting around 1,100 nm

with shorter wavelengths follow, from which again some have similar wavelengths and intensity distributions and build groups of modes. As one can see from Fig. 2.10 each two modes of the type EH_{mn} and $HE_{m+2,n}$ belong to a single group, for example $EH_{11}$and HE_{31}or EH_{12} and HE_{32}. Also modes of the type TE_{0n}, TM_{0n} and HE_{2n} belong together. HE-modes with the angular number 1 stay alone, like HE_{11}, HE_{12} or HE_{13}. Because only limited number of the higher order modes was calculated, not every group is shown complete. The threshold material gain increases for higher order modes because the field distribution moves to the outer regions and the overlap with the pumped active region decreases.

In Fig. 2.11 field profiles in a transverse plane inside of the cavity for the ground and higher order modes from the Fig. 2.10 are shown. Different symmetries can be observed by taking a look at the field profiles of different modes and comparing modes with the same angular number m as well as modes with different angular number but with the same sequence number n. The ground mode HE_{11} has a Gaussian-like profile and is linear polarized.

It is also of a great importance to know the radial intensity distribution for the ground and the higher order modes, which is shown in Figs. 2.12 and 2.13.

For higher order modes the intensity distribution moves from the center to outer regions, although this trend is not strongly regular. Lasing modes, which belong to the same group, have similar radial intensity distribution, for example TE_{01}, TM_{01} and HE_{21}. To get the better overview over the three-dimensional field distributions, electric field amplitudes for the ground mode HE_{11} and the first order mode TE_{01} for the same VCSEL are shown in Fig. 2.14.

The cylindrical symmetry of oxide-confined VCSELs in the reality could be disturbed during the fabrication, for example during the formation of the oxide aperture by the wet oxidation, depending on the oxidation parameters. Commonly, for large enough aperture diameters this effect becomes small, so that the mode structure remains similar to the case investigated here. Of course it could be desirable in some cases not to have the cylindrical symmetry, for example for polarization stabilization. Breaking the symmetry can be easily achieved by fabricating VCSELs with square mesas or by using other methods. In any case having a deeper understanding of the mode structure and the corresponding field profiles of the ground and the higher order modes enables to design optical properties of the VCSELs more efficiently.

2.1.4 Calculation of the Lasing Mode Parameters

By a known field distribution of a lasing mode important optical parameters can be calculated, which then can be used for device characterization and optimization. Decisive optical mode parameters are the effective mode volume V_{eff} and the three-dimensional optical confinement factor Γ. Both quantities play a major role for CW and high frequency operation of a semiconductor laser. The effective

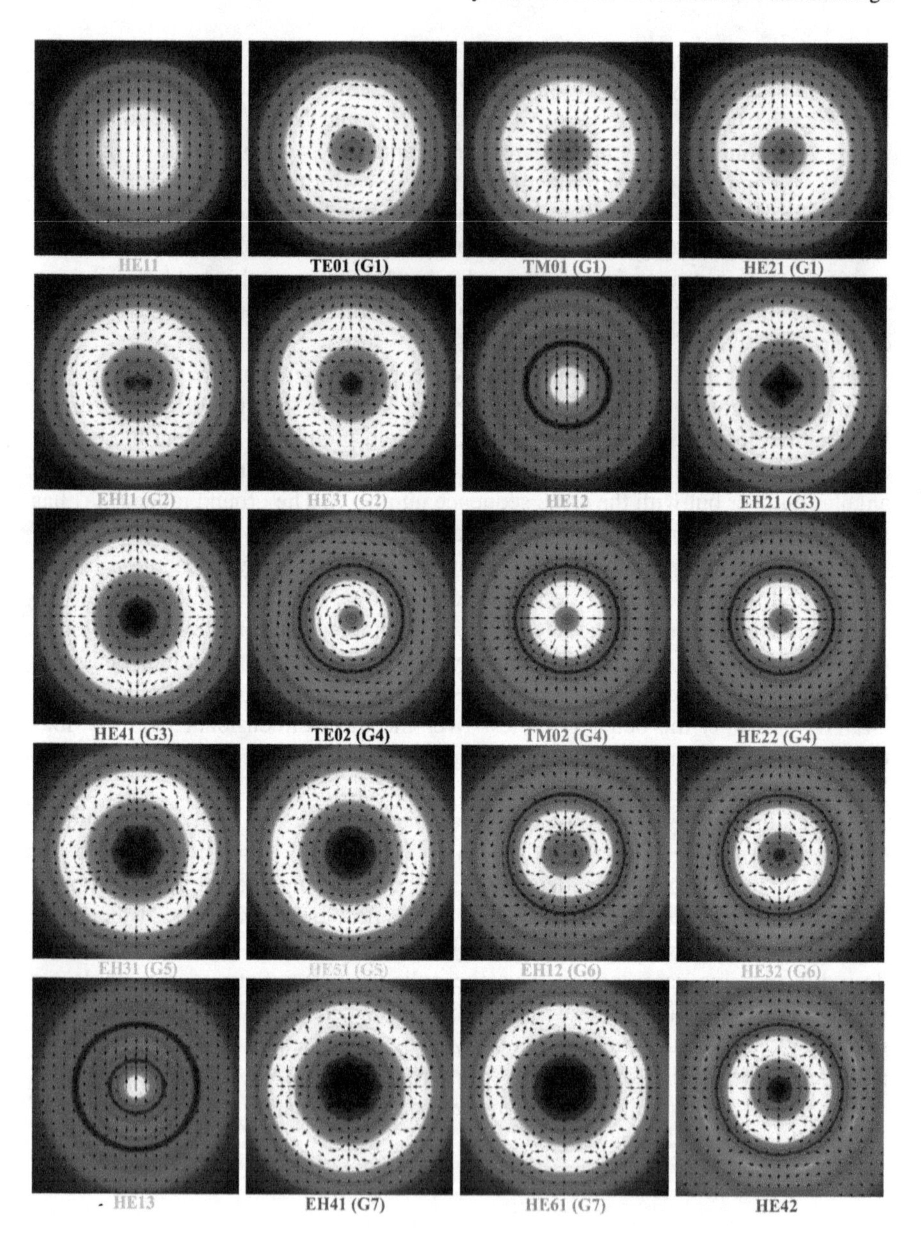

Fig. 2.11 Electrical field distribution in a transverse plane inside of the VCSEL cavity for 8 μm aperture diameter; the dimensions of each figure are 10 × 10 μm, the symmetry axis (z-axis) goes through the central point of each figure out of the figure plane, in brackets the group number is indicated

mode volume for each lasing mode can be calculated from the field distribution of this mode and the refractive index profile of the simulated structure using the formula [34]

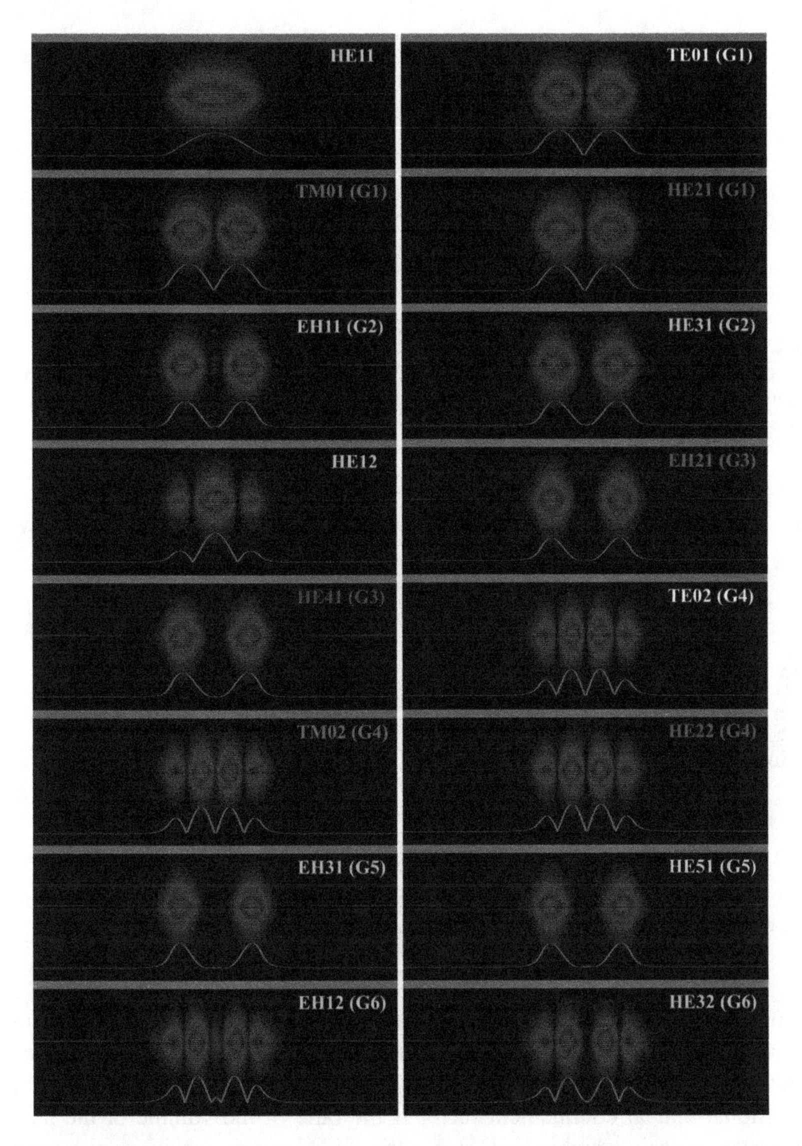

Fig. 2.12 Radial intensity profiles in the r–z-plane and radial electric field amplitudes (*red lines*) of the ground and the higher order modes for the VCSEL with 8 μm oxide aperture diameter; the width of each figure is 32 μm, the symmetry axis (z-axis) goes vertically across the center of each figure

$$V_{\text{eff}} = \frac{\iiint_{\text{cavity}} \varepsilon(\vec{r}) \left| \vec{E}(\vec{r}) \right|^2 d^3\vec{r}}{\max\left(\varepsilon(\vec{r}) \left| \vec{E}(\vec{r}) \right|^2 \right)}. \tag{2.1.44}$$

As can be obtained from Eq. 2.1.44 the effective mode volume is the volume of the cavity weighted by the electrical field power density, which is directly

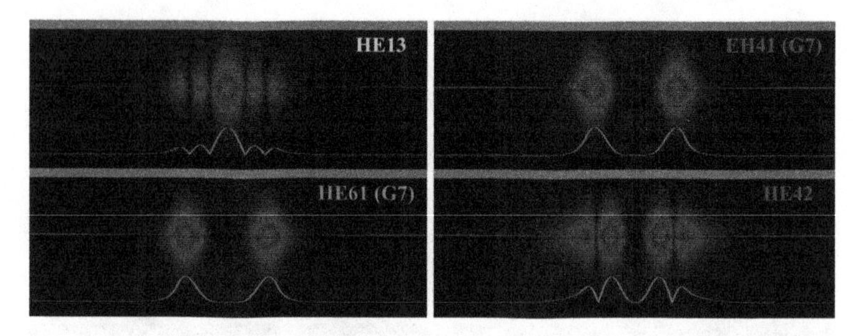

Fig. 2.13 Radial intensity profiles in the r–z-plane and radial and electric field amplitudes (*red lines*) of the higher order modes for the VCSEL with 8 μm oxide aperture diameter; the width of each figure is 32 μm, the symmetry axis (z-axis) goes vertically across the center of each figure

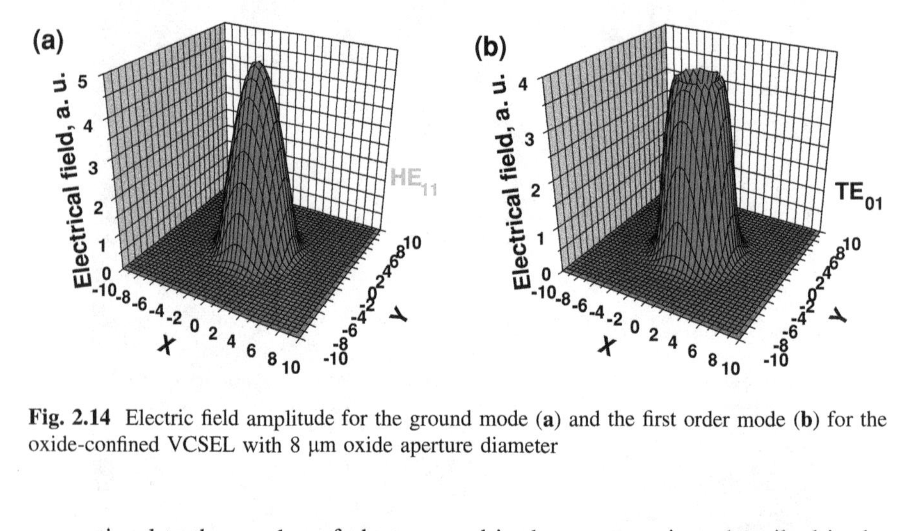

Fig. 2.14 Electric field amplitude for the ground mode (**a**) and the first order mode (**b**) for the oxide-confined VCSEL with 8 μm oxide aperture diameter

proportional to the number of photons used in the rate equations described in the following sections. As we will see later, the effective volume enters equations characterizing the high speed properties of directly modulated lasers. The three-dimensional optical confinement factor is the ratio of the volume of the pumped active region, again weighted by the electrical field power density, to the effective mode volume, as given by the following equation

$$\Gamma = \frac{\iiint_{\text{active}} \varepsilon(\vec{r}) \left|\vec{E}(\vec{r})\right|^2 d^3\vec{r}}{\iiint_{\text{cavity}} \varepsilon(\vec{r}) \left|\vec{E}(\vec{r})\right|^2 d^3\vec{r}}. \tag{2.1.45}$$

Additional to the three-dimensional optical confinement factor Γ also one-dimensional confinement factor Γ_Z can be calculated using one-dimensional methods, for example the transfer matrix method considered above:

$$\Gamma_Z = \frac{\int_{\text{active}} \varepsilon(z)\left|\vec{E}(z)\right|^2 dz}{\int_{\text{cavity}} \varepsilon(z)\left|\vec{E}(z)\right|^2 dz}. \tag{2.1.46}$$

The one-dimensional confinement factor Γ_Z represents commonly the best case value which the three-dimensional confinement factor Γ can achieve in a VCSEL, because of the additional edge effects in the three-dimensional case.

Very helpful for the understanding of the optical properties of VCSELs is a qualitative dependence of these parameters on the oxide aperture diameter. For this purpose, simulations on two simple 850 nm VCSEL structures with 24/35 top/bottom $Al_{0.15}Ga_{0.85}As/Al_{0.90}Ga_{0.10}As$ DBR pairs without any gradings were carried out. To investigate the dependence of the effective mode volume and the optical confinement factor on the cavity length one of the simulated structure was chosen to have a 1λ AlGaAs based cavity, the second structure had a 2λ cavity. For comparison of the three-dimensional and one-dimensional confinement factors and also of the wavelengths one-dimensional simulations with the transfer matrix method were carried out as well.

In Fig. 2.15 wavelength and optical confinement factors are given for the ground mode HE_{11} and the first order mode TE_{01}. For comparison one-dimensional values are also shown by blue lines and corresponding numbers.

From the figure one can see that the three-dimensional values both for the wavelength and for the optical confinement factor are smaller than the corresponding one-dimensional numbers. With increasing aperture diameters the three-dimensional values come closer to the one-dimensional case, because the fields become more similar to the plane wave approximation. Another issue is that for a homogeneously pumped active region with the diameter of the pumped zone 2 μm larger than aperture, like in the case of the presented simulations, the optical confinement factor of the first order mode is smaller than for the ground mode. This can be explained by the field distribution of both modes. Optical energy is

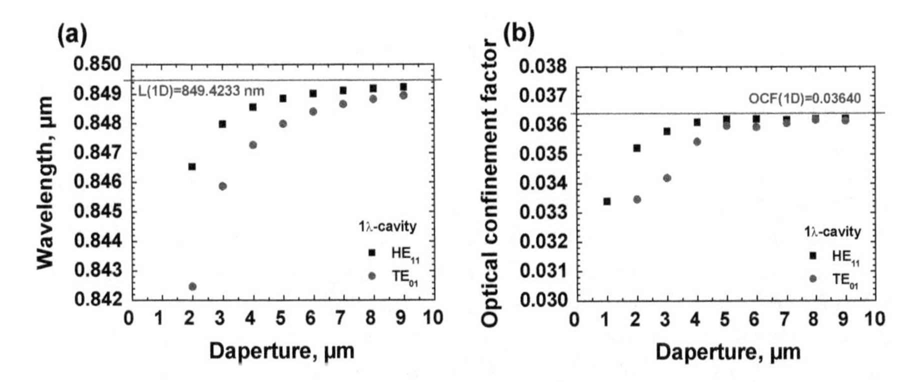

Fig. 2.15 Wavelength (**a**) and optical confinement factor (**b**) for the ground mode (*black squares*) and the first order mode (*red circles*) as a function of the aperture diameter for the 1λ cavity VCSEL; *blue lines* and *numbers* represent one-dimensional values

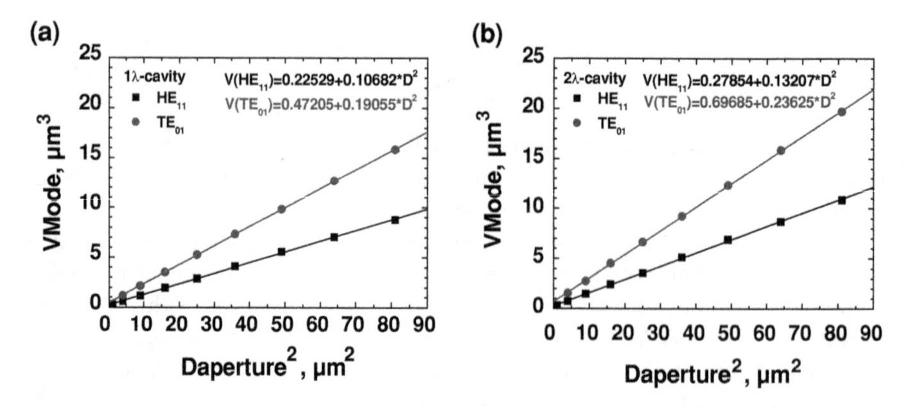

Fig. 2.16 Effective mode volume for the 1λ cavity VCSEL (**a**) and for the 2λ cavity VCSEL (**b**) for the ground and the first order modes as a function of the squared aperture diameter, together with linear fits and corresponding formulas

concentrated for the HE_{11} mode closer to the center as for the TE_{01} mode (see Fig. 2.11 and Fig. 2.12). Very interesting is also the dependence of the effective mode volume on the aperture diameter for both modes, which is shown in Fig. 2.16.

One can see that both modes have linear dependence of the effective mode volume on the squared aperture diameter, i.e. on the aperture area. An offset for both effective mode volumes for both devices is present, as can be obtained from the formulas shown on the figures. It can be explained by the fact that a part of the optical energy is located outside of the aperture region, so that at zero aperture diameters this part would contribute to the mode volume. The smallest aperture diameter investigated was 1 μm and it fits also to the linear dependence. The offset is logically larger for the TE_{01} mode, because this mode has more energy located in the outside regions. Mode volumes for the TE_{01} are also larger compared to the HE_{11} mode, because the first order mode is weaker confined. The linear dependence of the mode volume on the aperture diameter can be experimentally proved by measurement of the parameters which depend on the effective mode volume, for example the *D*-factor. These measurements will be described in the later chapters.

The slope of both curves is larger for the longer VCSEL, because of the larger cavity length. The slope of the effective mode volume for the HE_{11} mode is 0.10682 for the 1λ cavity and 0.13207 for the 2λ cavity. The effective cavity length calculated with the transfer matrix method is 0.4783 μm for the shorter device and 0.5917 μm for the longer one. The ratio of the slopes of the linear fits for the ground mode 0.13207/0.10682 is 1.236 and practically perfectly agrees to the ratio of the effective cavity lengths of 0.5917/0.4783, which is 1.237. This fact shows that the differences in the slope of the effective mode volume curves of the ground mode are mainly caused by the differences in the effective cavity lengths of the VCSELs.

Thus one can see that one-dimensional and three-dimensional simulations are powerful tools to deliver the knowledge about important parameters of the lasing modes. This helps by the characterization, optimization and understanding of the optical phenomena inside of VCSELs, since optical processes in the devices are complicated and versatile. Nevertheless optics is only one of the many sides of semiconductor lasers, and in the next sections electrical and thermal phenomena will be considered.

2.2 Electrical Properties

Semiconductor lasers are devices which convert electrical energy in the optical energy, thus optoelectronic devices. This means that additional to the optical phenomena considered in the previous sections, also electrical phenomena play a tremendous role in the semiconductor lasers. As in the case of the optical properties, electrical properties are also determined by the design of both epitaxial structure and of the fabricated device. While in the case of the optical properties in oxide-confined VCSELs the main influence during the fabrication comes from the formation of the oxide aperture, more effects contribute to the electrical properties of the fabricated devices, especially on the high frequency side. Such parameters, which are determined by the device design, like dimensions and position of the contacts, dimensions of the contact pads, properties of the applied dielectric layers, play a decisive role for the electrical properties of the lasers. That is why it makes sense to distinguish between the electrical properties, which should be considered during the designing of the epitaxial growth structure, and external electrical properties of the fabricated devices, which act mostly as a limitation of the internal electrical properties. In the next three sections we will describe designing of the internal electrical properties of the epitaxial structure, while in the following two sections external effects will be considered. During the designing process of any semiconductor lasers with active elements of dimensions, where quantum effects play a significant role, for calculation of the corresponding properties quantum mechanics should be applied. This is the fact for QW-regions, and thus we will consider quantum mechanic effects also in the following sections. In the sections handling the electrical properties we will not use the phasor notation, so that all quantities are real physically values.

2.2.1 Carrier Dynamic in Semiconductor Lasers

We start again with the Maxwell's equations (2.1.1–2.1.4) [1, 35]. For quasi-electrostatic fields, that means for fields with slow time variation, one can neglect the time derivatives in Eqs. 2.1.1 and 2.1.2 and for the case of absent or negligible magnetic field ($\vec{H} \cong 0$, $\vec{B} \cong 0$) for electric field two equations will stay:

$$\vec{\nabla} \times \vec{E} = 0, \tag{2.2.1}$$

$$\vec{\nabla} \cdot \vec{D} = \rho. \tag{2.2.2}$$

The solution for the electric field can be expressed in this case by a gradient of a scalar function φ, which is called electrostatic potential:

$$\vec{E} = -\vec{\nabla}\varphi, \tag{2.2.3}$$

Inserting (2.2.3) into (2.2.2) and using (2.1.6) for an isotropic medium we get the Poisson's equation for the electrostatic potential:

$$\vec{\nabla} \cdot \left(\varepsilon\vec{\nabla}\varphi\right) = -\rho. \tag{2.2.4}$$

By higher frequencies one may include the displacement current density $\vec{J}_{\text{disp}}$ to the conduction current density, so that for the total current density $\vec{J}_{\text{tot}}$ the following equation holds:

$$\vec{J}_{\text{tot}} = \vec{J}_{\text{con}} + \vec{J}_{\text{disp}} = \vec{J}_{\text{con}} + \frac{\partial}{\partial t}\left(\varepsilon\vec{E}\right). \tag{2.2.5}$$

The charge density ρ can be expressed in a semiconductor by a sum over the electron concentration n, the hole concentration p and concentrations of the positive charged ionized donors N_D^+ and the negative charged ionized acceptors N_A^- with appropriate electrical charge sign:

$$\rho = q\left(p - n + N_D^+ - N_A^-\right), \tag{2.2.6}$$

where q stay for the magnitude of the unit charge and is $q = 1.6 \times 10^{19}$ C. One can divide the conduction current density into two parts representing electron and hole conduction current densities $\vec{J}_n$ and $\vec{J}_p$ respectively:

$$\vec{J}_{\text{con}} = \vec{J}_p + \vec{J}_n. \tag{2.2.7}$$

By a time-independent total doping concentration $C_0 = N_D^+ - N_A^-$ one can divide the continuity equation (2.1.5) for the conduction current into two parts for electrons and holes:

$$\vec{\nabla} \cdot \vec{J}_n - q\frac{\partial}{\partial t}n = qR, \tag{2.2.8}$$

$$\vec{\nabla} \cdot \vec{J}_p + q\frac{\partial}{\partial t}p = -qR, \tag{2.2.9}$$

with the net recombination rate of electron–hole pairs per unit volume R given by

$$R = R_n - G_n = R_p - G_p \tag{2.2.10}$$

with the generation (G_n and G_p) and recombination (R_n and R_p) rates for electron and holes separately. Thus for each kind of the charge carriers continuity equation can be written:

$$\frac{\partial}{\partial t}n = G_n - R_n + \frac{1}{q}\vec{\nabla} \cdot \vec{J}_n, \quad (2.2.11)$$

$$\frac{\partial}{\partial t}p = G_p - R_p - \frac{1}{q}\vec{\nabla} \cdot \vec{J}_p. \quad (2.2.12)$$

For the flow of charge carriers drift–diffusion model can be used, which consider two major mechanisms of the carrier transport. The first is the drift current which is generated by an electrical field and is directly proportional to the conductivity of electrons $\sigma_n = q\mu_n n$ and holes $\sigma_p = q\mu_p n$, where μ_n and μ_p are the electron and hole mobility, respectively. By a present concentration gradient of electron or holes diffusion current will appear, which is directly proportional to the diffusion coefficient of the electrons D_n and holes D_p. Mobility and diffusion coefficients are related by the Einstein relation

$$\frac{D_n}{\mu_n} = \frac{D_p}{\mu_p} = \frac{kT}{q}, \quad (2.2.13)$$

where T is the temperature of the material and k is the Boltzmann constant $k = 1.3807 \times 10^{-23}$ J/K. The total current density of electron and holes can be then written as

$$\vec{J}_n = q\mu_n n\vec{E} + qD_n \nabla n, \quad (2.2.14)$$

$$\vec{J}_p = q\mu_p p\vec{E} - qD_p \nabla p, \quad (2.2.15)$$

Inserting (2.2.14) and (2.2.15) into (2.2.11) and (2.2.12) and using electrostatic potential (2.2.3) instead of the electric field allows us to reduce the number of unknowns to only three scalar functions $n(\vec{r})$, $p(\vec{r})$ and $\varphi(\vec{r})$:

$$\frac{\partial}{\partial t}n = G_n - R_n + \frac{1}{q}\vec{\nabla} \cdot \left(q\mu_n n\vec{\nabla}\varphi + qD_n\vec{\nabla}n\right), \quad (2.2.16)$$

$$\frac{\partial}{\partial t}p = G_p - R_p - \frac{1}{q}\vec{\nabla} \cdot \left(q\mu_p p\vec{\nabla}\varphi - qD_p\vec{\nabla}p\right), \quad (2.2.17)$$

$$\vec{\nabla} \cdot \left(\varepsilon\vec{\nabla}\varphi\right) = -q\left(p - n + N_D^+ - N_A^-\right). \quad (2.2.18)$$

By specified boundary conditions Eqs. 2.2.16–2.2.18 can be solved and will deliver the three unknowns. Inside of the laser the carrier concentrations can vary by many orders of magnitude. That is why they are often replaced by the quasi-Fermi levels F_n and F_p for electrons and holes, respectively. They can be calculated in a common case using the Fermi–Dirac statistics for electrons f_n and holes f_p, the corresponding densities of states ρ_e and ρ_h and the following equations:

$$n = \int_{-\infty}^{\infty} f_n(E)\rho_e(E)dE, \tag{2.2.19}$$

$$p = \int_{-\infty}^{\infty} f_p(E)\rho_h(E)dE, \tag{2.2.20}$$

with

$$f_n(E) = \frac{1}{1 + e^{(E-F_n)/kT}}, \tag{2.2.21}$$

$$f_p(E) = \frac{1}{1 + e^{(F_p-E)/kT}}. \tag{2.2.22}$$

One can express the quasi-Fermi levels for electrons F_n and holes F_p by corresponding potentials according to the following equations:

$$F_n(\vec{r}) = -q\varphi_n(\vec{r}) + E_{ref}, \tag{2.2.23}$$

$$F_p(\vec{r}) = -q\varphi_p(\vec{r}) + E_{ref}, \tag{2.2.24}$$

with a reference constant energy E_{ref} and handle the three unknowns $\varphi(\vec{r})$, $\varphi_n(\vec{r})$ and $\varphi_p(\vec{r})$, which have the same units (volts).

Using the drift–diffusion model one can efficiently design epitaxial structure of a VCSEL and optimize DBR layers and the cavity region. For designing of the active region quantum mechanics should be applied. In the next two sections one-dimensional calculations for the epitaxial VCSEL design optimization, carried out with the free software Nextnano++ [36], are presented. The used software includes a solver for the Poisson's and current equations and has quantum mechanical solver for the active region as well. It is simple to use and is very fast, thus perfectly suited for such kinds of applications.

2.2.2 Electrical Design of the DBR Mirrors

The largest part of the epitaxial VCSEL structure consists of DBR mirrors, both top and bottom. DBR mirrors contain also the most number of interfaces of a VCSEL structure. That is why a proper and careful electrical design of both mirrors is of a great importance. Since the thicknesses of the DBR-layers are mainly defined by the wavelength of the laser light, the main objective of electrical simulations is to improve the conductivity of the mirrors, mainly by reducing potential barriers at the interfaces, which arise from the different material compositions and thus different band gaps of the corresponding materials used in the DBR mirrors. For the VCSELs presented in this work Al-compositions of 12, 15 and 90% additional to pure GaAs were used. We will describe optimization of the DBR mirror structure on the example of the 980 nm QW-VCSELs investigated in this work.

Immense work was done in the past for optimization of the interfaces inside a DBR mirror [37]. Different schemes were investigated, for example linear gradings (LG) at the interfaces or bi-parabolic gradings. While the second scheme is more difficult to realize, linear gradings can be grown without big effort by standard metal–organic chemical vapor deposition (MOCVD) technique. In the case of the molecular beam epitaxy (MBE) digital gradings should be applied, but it can be realized as well. In this work, for all structures linear gradings of 10 or 20 nm thickness for both top and bottom DBR mirrors were applied.

In Fig. 2.17 Al-composition for an $Al_{0.12}Ga_{0.88}As/Al_{0.90}Ga_{0.10}As$ DBR mirror without and with 20 nm thick linear gradings is presented. The total thickness of each DBR pair is kept constant. The corresponding band structures are shown in Fig. 2.18 for the case without gradings (a) and with the 20 nm thick gradings (b) for intrinsic material. In the following pictures only two periods of the DBR mirrors are shown for more clarity.

In Fig. 2.18 energies of different minima and maxima in conduction and valence bands are shown. Electrons can occupy the minima around the Γ, X and L points in the conduction band. Holes occupy the maxima in the heavy-hole (HH), light-hole (LH) or split-off (SO) valence bands. For Al-compositions from 0 up to ~40–45% AlGaAs has a direct band gap, for higher compositions AlGaAs becomes an indirect semiconductor with the X minima having the lowest energy in the conduction band.

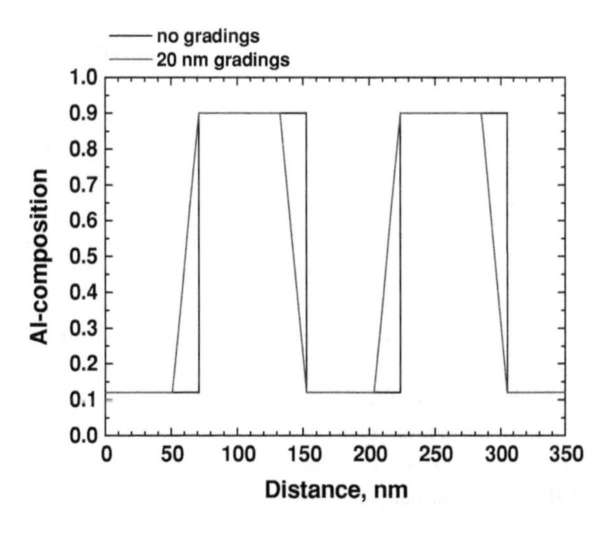

Fig. 2.17 Al-composition of an $Al_{0.12}Ga_{0.88}As/Al_{0.90}Ga_{0.10}As$ DBR mirror without and with 20 nm thick linear gradings

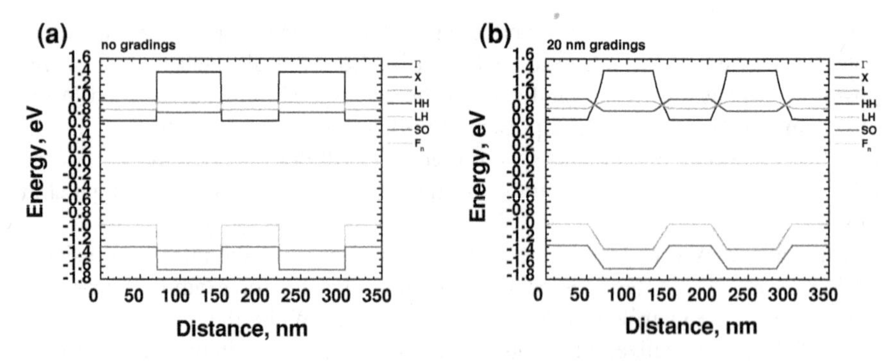

Fig. 2.18 Band structure of the DBR mirror without (**a**) and with 20 nm thick linear gradings (**b**) for intrinsic materials

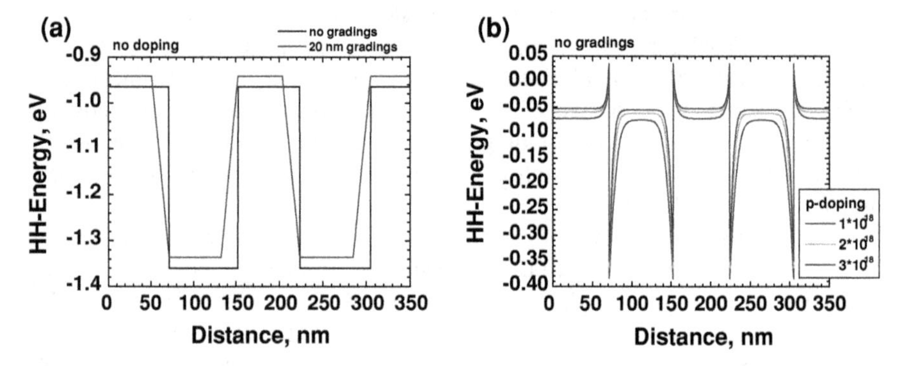

Fig. 2.19 Energy of the heavy-hole band minimum for the DBR mirror with and without linear gradings for intrinsic semiconductor (**a**) and for the DBR mirror without gradings for different levels of the p-doping (**b**)

From the Fig. 2.18 one can see considerable potential barriers between materials with different Al-composition for both electrons in the conduction band and holes in the valence band. In Fig. 2.19a, the zoomed view of the heavy-hole band minimum for intrinsic materials for the cases without and with linear gradings is shown. The height of the potential barrier for holes reaches ~0.4 eV. If one considers the total number of the interfaces in a typical DBR mirror, which varies in the range of 40–70 for 20 and 35 DBR pairs respectively, one can immediately see the importance of reducing the barriers. The commonly used solution is to dope the corresponding mirror whether with donors for the n-type DBR or acceptors for the p-type mirror. In the Fig. 2.19b energy of the heavy-hole band minimum is shown for the case of different p-type doping levels of 1×10^{18}, 2×10^{18}, and 3×10^{18} cm^{-3}, which are typical doping levels for DBR mirrors.

With the present p-doping the situation with the interfaces improves drastically. The difference in the heavy-hole band minimum energies for materials with different composition practically vanishes, and only very thin barriers stay.

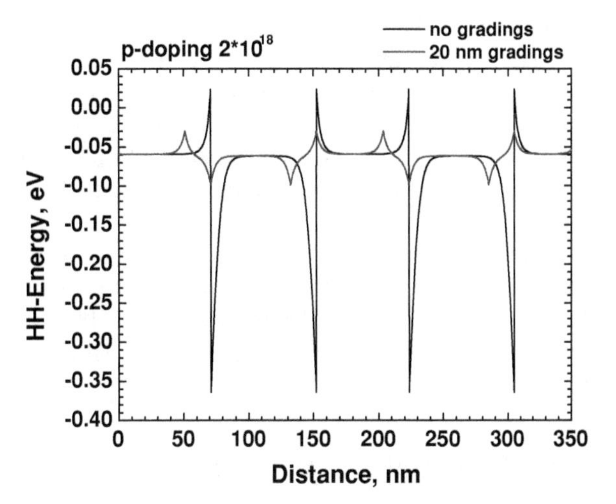

Fig. 2.20 Energy of the heavy-hole band minimum for the p-doping level of 2×10^{18} cm^{-3} for the DBR mirror without and with 20 nm thick linear gradings

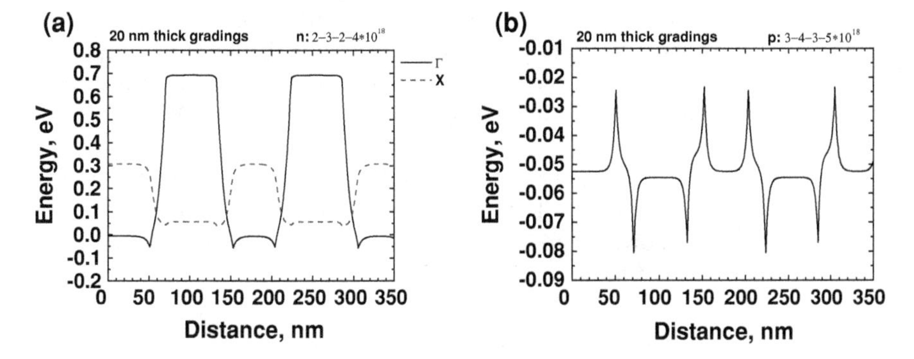

Fig. 2.21 Energy of the Γ and X points of the conduction band for the n-doped DBR mirror (**a**) and energy of the heavy-hole band minimum for the p-doped DBR mirror (**b**)

The thickness of the residual barriers depends on the doping level. Hereby the difference between the doping levels of 2×10^{18} and 3×10^{18} cm^{-3} become smaller as compared to the difference between 1×10^{18} and 2×10^{18} cm^{-3}. Remaining barriers are reduces further for the DBR mirrors with the linear gradings, as shown in Fig. 2.20 for the p-doped mirror and Fig. 2.21 for the n-doped mirror.

The barrier height is reduced from ~0.38 eV down to ~0.07 eV for one p-doped interface. After consideration of the simulation results, doping concentrations of 2×10^{18} cm^{-3} and higher for both n-type and p-type DBR mirrors were chosen for the 980 nm QW-VCSELs. These doping levels are sufficient to reduce the barriers and from the other side do not increase absorption losses in a drastically way. In the first few DBR mirror pairs close to the cavity the doping was leaved at 2×10^{18} cm^{-3}, while moving further away from the cavity into the region, where the photon density decreases, enables to use higher doping levels. Hereby for interfaces, where electrical field has a node, higher doping levels can be used as for interfaces, in which electrical field has an anti-node. For the investigated structure n-doping levels following the scheme 2–3 to 2–4 × 10^{18} cm^{-3}, for the corresponding composition scheme $Al_{0.12}Ga_{0.88}As$-LG-$Al_{0.90}Ga_{0.10}As$-LG, was applied. For the p-type DBR the doping levels were 3–4 to 3–5 × 10^{18} cm^{-3}.

Simulation results presented in this section show, that by introducing linear gradings into the DBR mirrors potential barriers can be effectively suppressed. Linear gradings are an effective and relatively easy to realize solution for low resistance multilayered mirrors for both n- and p-type DBRs.

2.2.3 Design of the Active Region

The part of a semiconductor laser, in which electrical power is converted to the optical power, is the active region. Physical properties of the active region determine in a decisive way the properties of the laser, both for CW and high frequency modulation. That is why special attention must be paid while designing the active region. We will describe this again using the 980 nm QW-VCSELs as example.

Quantum mechanics should be applied in order to be able to describe physical properties of QWs correctly. The basic nonrelativistic quantum mechanical equation is the well-known Schrödinger equation [1, 2, 35], which for a static case and one particle has the form

$$H\Psi = E\Psi, \tag{2.2.25}$$

with

$$H = -\frac{\hbar^2}{2m}\nabla^2 + V(\vec{r}). \tag{2.2.26}$$

Here H is the quantum mechanical operator called Hamiltonian, $\hbar$ is the Planck constant h divided by 2π, and m is the mass of the desired particle. The solution of the Schrödinger equation for some energy value E is the corresponding Ψ-function, also called the wave function, describing all physical properties of the quantum mechanical state of the particle. $V(\vec{r})$ is the potential energy function, which defines the particle motion.

In the case of semiconductor lasers considered particles are electrons and holes. For a simpler description we can use in the first approximation energy of the corresponding minima and maxima in the conduction and valence bands as potential energy and so-called effective masses of the carriers instead of the masses of free particles. This enables to use the simple Eqs. 2.2.25 and 2.2.26 and at the same time to include effects corresponding to the band structure. Also more advanced models like 8-band k*p theory [1] can be applied for more exact calculations. Nextnano++ is able to calculate energy states using both classical method using effective carrier masses and the 8-band k*p method.

The most important property which should be calculated is the emission wavelength of the laser. In the case of a VCSEL the wavelength of the emitted light is defined by the cavity dip position, thus by the optical properties of the cavity, described in the previous sections. Nevertheless the active medium used in the VCSEL should be able to emit photons at this wavelength; otherwise no lasing would take place. Because of the temperature effects, the peak of the stimulated emission in the active region should be designed to be at a shorter wavelength compared to the cavity dip wavelength. The reason is the stronger temperature shift of the band gap of AlGaAs compared to the shift of the wavelength of the cavity dip. Because internal temperatures in VCSELs can easily reach 80–100°C, detuning between the cavity dip wavelength and the peak emission wavelength of the active region is commonly in the range of 5–20 nm. For a stable high temperature operation rather larger values should be considered. From 980 nm QW-VCSELs described in this work nominal value of the cavity-gain detuning was 15 nm, thus the peak emission wavelength of the active region was calculated to be at 965 nm. At this wavelength strained InGaAs QWs can be used. After some calculations the In-composition of 21% by the QW-thickness of 4.2 nm was determined. For strain compensation 6 nm thick GaAsP layers acting at the same time as barriers between QWs with P-composition of 12% were applied. Figure 2.22 shows the corresponding bands for the active region considered here. Because of the strain the HH and LH valence bands are splitted.

By solving the Schrödinger equation with the effective masses of the corresponding band minima and maxima eigenvalues of the energy states and corresponding Ψ-functions were calculated. Hereby only the first five solutions for each electrons and holes, together ten solutions, were investigated. They correspond to the ground levels in the conduction and valence bands. Figure 2.23 shows the calculated energy levels for electrons (a) and holes (b).

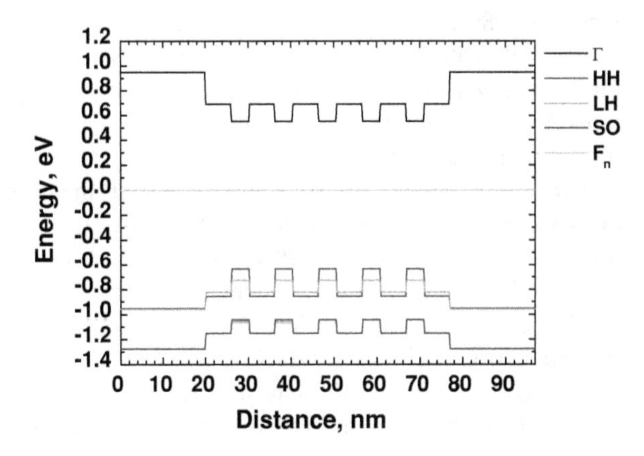

Fig. 2.22 Energy diagram of the active region consisting of five 4.2 nm thick $In_{0.21}Ga_{0.79}As$ QWs with 6 nm thick $GaAs_{0.88}P_{0.12}$ barriers acting simultaneously as strain compensation layers; the outer region consist on each side of 20 nm $Al_{0.35}Ga_{0.65}As$

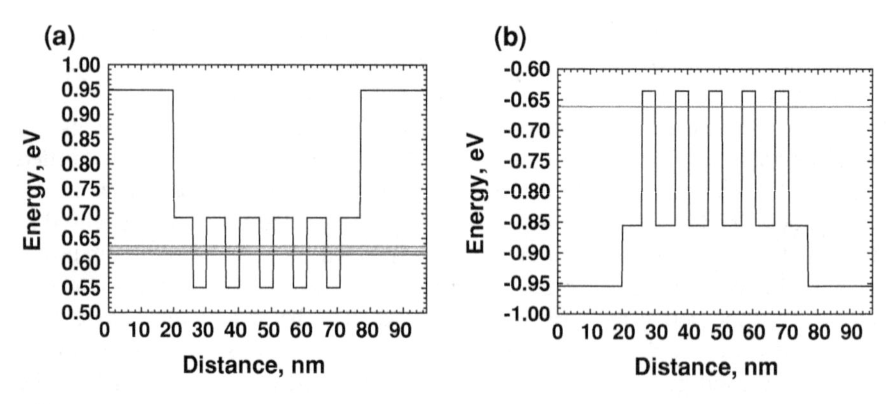

Fig. 2.23 Energy states in the conduction (**a**) and the heavy-hole (**b**) band for electrons and holes respectively

The QWs are partially coupled, as can be seen from the energy levels, which has slightly different energy values. Energy levels of the hole states have practically identical values. Figure 2.24 shows the first four energy levels with the corresponding wave functions for the electrons in the conduction band. Corresponding hole states and wave functions are shown in Fig. 2.25.

As one can see from the Figs. 2.24 and 2.25, the form of the corresponding wave functions of electrons and holes is similar and their overlap is large, resulting in an efficient stimulated photon emission.

Finally the difference between the lowest electron energy level and the highest hole energy level should be calculated in order to determine the emission wavelength. In Fig. 2.26 the energy difference between the ground electron and hole levels as a function of the temperature, calculated with the effective mass theory (black) and with the more precisely 8-band k*p theory (red), is shown. For a better orientation the goal wavelength of the active region peak emission at room temperature of 965 nm (green line), the wavelength of the room temperature VCSEL emission determined by the cavity dip wavelength at 980 nm (blue dotted line) and the temperature dependent cavity dip wavelength (blue solid line) are also shown. Considered was a temperature range between 300 and 400 K, which corresponds approximately to temperatures from the room temperature up to ~120°C. Since the cavity dip also moves with the temperature, the best resonance between the active region peak gain and the cavity dip will appear not at the internal VCSEL temperature of ~345 K, corresponding to ~70°C, but at a higher temperature. Since internal temperatures of ~70°C are reached at the ambience temperatures of ~20–30°C, for a better temperature stability and an efficient operation at ambience temperatures of 80–100°C, which result in internal laser temperatures of 130–150°C, or ~400–420 K, perfect detuning should be designed to lay somewhere between 340 and 400 K. Assuming an approximately fourfold weaker dependence of the cavity dip wavelength on temperature, as compared

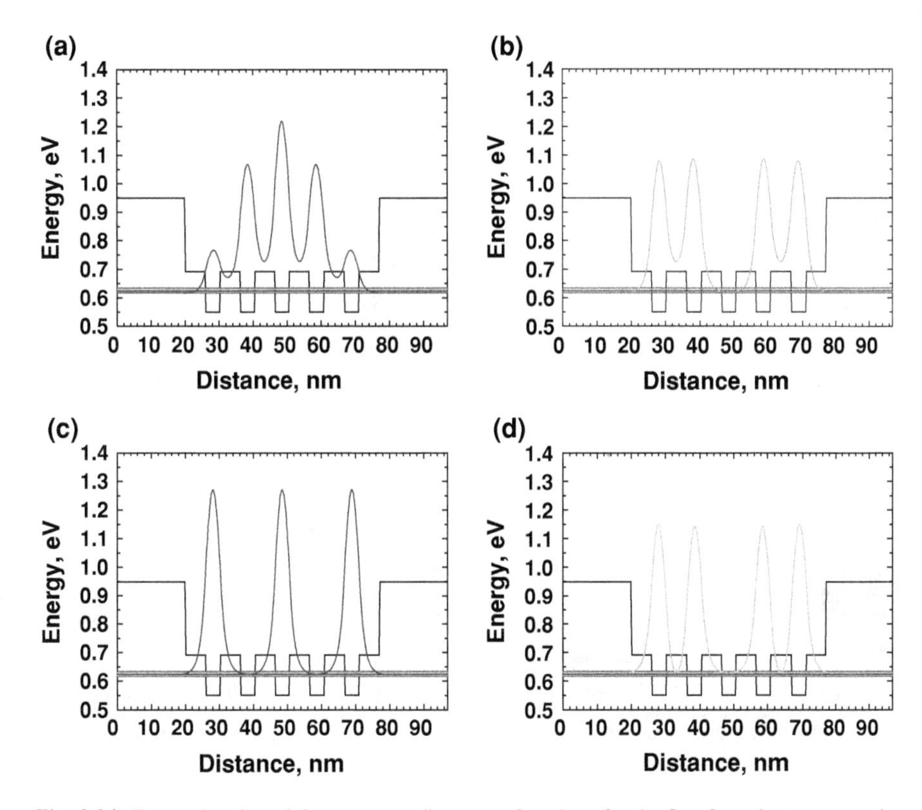

Fig. 2.24 Energy levels and the corresponding wave functions for the first four electron states in the conduction band in the active region

to the band gap temperature dependence, the perfect matching of the peak gain and the cavity wavelengths would take place at ~360 K, as can be obtained from the Fig. 2.26. This corresponds to the internal laser temperature of ~90°C, which will be achieved at outer temperatures in the range of 40–50°C. This enables stable laser operation in the whole desired temperature range, from 20°C up to 85–100°C.

Using quantum mechanical methods, e.g. effective mass theory and 8-band k*p theory, energy levels and wave functions in the QWs can be easily calculated. Consequently, temperature dependent emission wavelength of the active region can be efficiently designed. This information is inalienable for a correct VCSEL operation. But additionally to the internal electrical properties of the laser external limitations should be carefully considered. This includes electrical parasitics inside of the VCSEL as well as a proper design of the high frequency contact pads. Both of these effects should be designed to enable high speed VCSEL operation at desired data rates, which were for this work as high as 35–40 Gbit/s. These subjects will be considered in the next two sections.

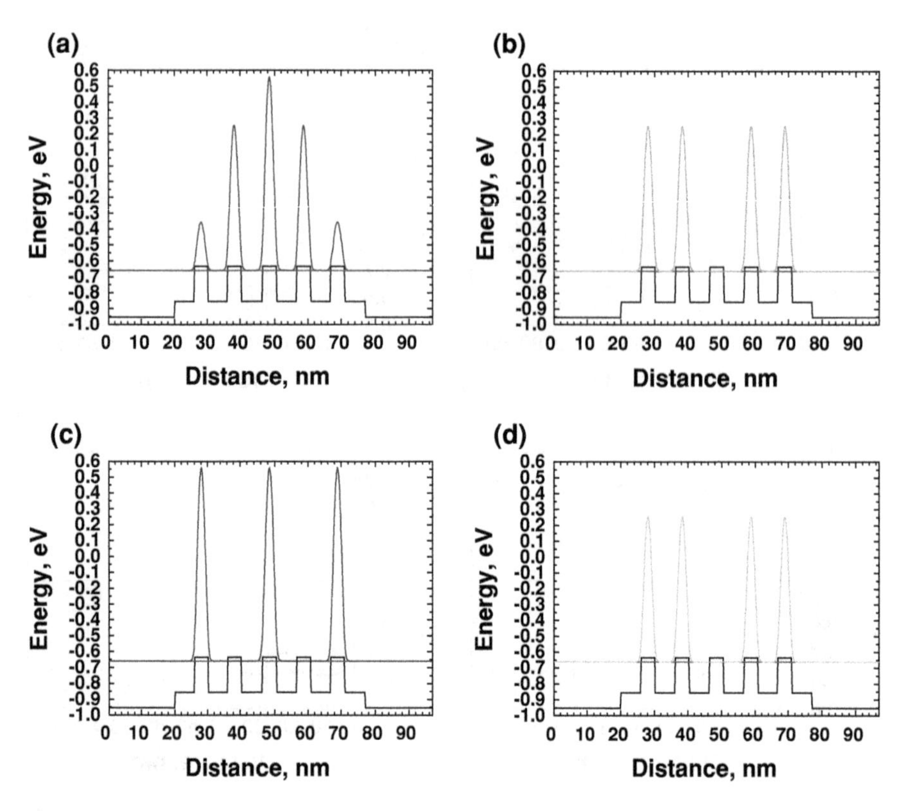

Fig. 2.25 Energy levels and the corresponding wave functions for the first four hole states in the heavy-hole band in the active region

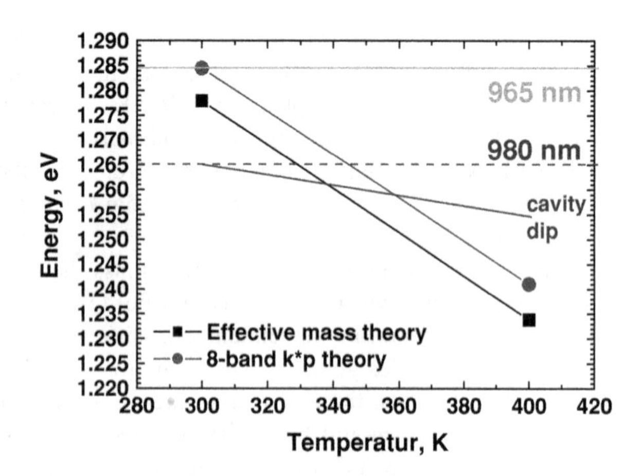

Fig. 2.26 Energy difference between the ground electron and hole levels as a function of the temperature, calculated with the effective mass theory and 8-band k*p theory, together with the goal wavelength of the active region peak emission at room temperature of 965 nm (*green line*), wavelength of the room temperature VCSEL emission at 980 nm (*blue dotted line*) and the temperature dependent cavity dip wavelength (*blue solid line*)

2.2.4 Equivalent Circuit and Electrical Parasitics of a VCSEL

For a proper operation of a semiconductor laser at high bit rates, high speed electrical current modulation driving the VCSEL should be transferred to the active region, where electro-optical conversion takes place, ideally without any distortions and attenuations. In a real VCSEL many effects prevent this. One of the most important of them is the presence of electrical parasitic elements, like unwanted capacitances and resistances, inside of the device, which build a low pass for the incoming high frequency electrical signal [38–41]. The exact form of the present electrical parasitic elements is determined first by the epitaxial structure, among other by doping levels, thickness and number of the oxide apertures etc., but also by the device design parameters chosen for fabrication, e.g. mesa size, oxide aperture diameter etc. That is why appropriate device design is of a great importance additionally to a proper epitaxial structure. In Fig. 2.27, a schematic picture of the electrical parasitic capacitances and resistances presented in a VCSEL is shown.

As one can see from Fig. 2.27a, additional to the resistance R_a and capacitance C_j, which represent the pumped active region, capacitance of the oxide aperture and the underlying intrinsic region C_o and of the contact pads C_p, as well as both resistances of the top and of the bottom DBR mirror R_{mt} and R_{mb}, should be considered. Since the two capacitances C_o and C_j are connected in parallel, they can be combined to one resistance C_a. According to the n-port theory [42], also both mirror resistances R_{mt} and R_{mb} together with eventually considerable contact resistances (not shown in the figure) can be represented by one common resistance R_m. By the given geometrical and structural parameters of a VCSEL all parasitic elements are defined and can be estimated using appropriate measurement techniques, for example S-parameter measurements [42].

The theory describing electrical high frequency properties of common network elements is the well known n-port theory [42]. Each port has two connections, and for VCSEL characterization the 2-port theory can be applied. The 2-port, describing the electrical parasitics in a VCSEL, is shown in Fig. 2.27b and is

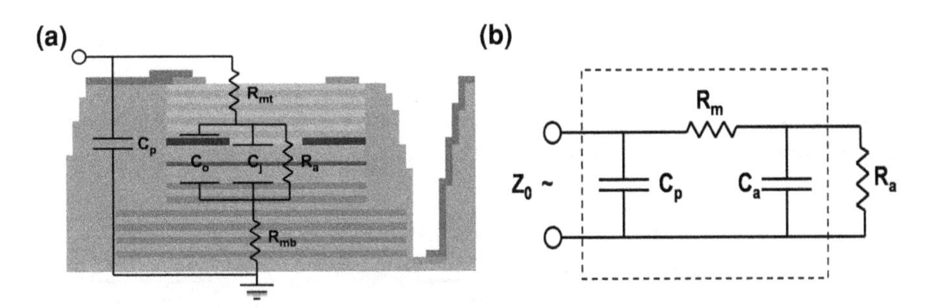

Fig. 2.27 Electrical parasitic capacitances and resistances inside a VCSEL (**a**) and the corresponding equivalent circuit with two parallel capacitances C_o and C_j combined together to C_a and resistances of the both mirrors R_{mt} and R_{mb} joined to R_m (**b**)

pointed by the dashed box. The first port connects a high frequency driving signal generator with impedance Z_0, commonly equal to 50 Ω, with the contact pads of the VCSEL. The second port describes the interface between electrical parasitics and the active region with the resistance R_a, which is the final destination of the incoming high frequency signal. All distortions and attenuations of the electrical signal within the dashed box are undesirable.

In Fig. 2.28 a common 2-port is shown. We describe the two ports symmetrically, so that a_i represents the incoming and b_i the outgoing signal for each port i. Z_i are reference impedances, which can be chosen free, but in the most cases it is comfortable to chose Z_1 to be equal to the generator (or driver) output impedance (commonly 50 Ω) and for Z_2 to use the resistance of the active region R_a of the desired VCSEL.

A 2-port can contain many elements, but its high frequency behavior is complete described by a 2 × 2 matrix, connecting the input and output signal magnitudes and phases. Hereby one can chose between several representations, which are free convertible. The most frequently used matrix is the S-matrix, or the so-called scattering matrix. The S-matrix handles not the voltage u_i and the current i_i at each port, but the forward propagating a_i and the backward propagating b_i waves. Forwards represents in this case the direction to the port and backwards the opposite direction. These waves for the two ports are shown in Fig. 2.28. The waves a_i and b_i are directly connected to the voltages u_i and currents i_i at the corresponding ports via the reference impedances Z_i, which are commonly real:

$$a_i = \frac{1}{2}\left(\frac{u_i}{\sqrt{Z_i}} + i_i\sqrt{Z_i}\right), \tag{2.2.27}$$

$$b_i = \frac{1}{2}\left(\frac{u_i}{\sqrt{Z_i}} - i_i\sqrt{Z_i}\right). \tag{2.2.28}$$

Corresponding voltages and currents on each port can be easily calculated from the a_i and b_i waves by using the reverse equations:

$$u_i = (a_i + b_i)\sqrt{Z_i}, \tag{2.2.29}$$

$$i_i = \frac{(a_i - b_i)}{\sqrt{Z_i}}. \tag{2.2.30}$$

Knowing the 2-port structure one can calculate the S-matrix, which connects the incoming and reflected waves to each other:

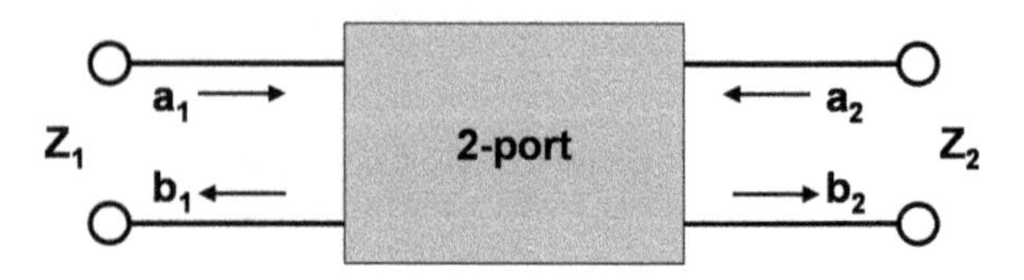

Fig. 2.28 A common 2-port with the symmetrical port notation

$$\begin{pmatrix} b_1 \\ b_2 \end{pmatrix} = S \cdot \begin{pmatrix} a_1 \\ a_2 \end{pmatrix} = \begin{pmatrix} S_{11} & S_{12} \\ S_{21} & S_{22} \end{pmatrix} \begin{pmatrix} a_1 \\ a_2 \end{pmatrix}. \tag{2.2.31}$$

The S-matrix includes additional to the parameters of the electrical elements of the 2-port also the reference impedances Z_i, that means it depends on some outer parameters, like generator impedance etc. There are several other matrices used in the 2-port, or more common n-port, theory, which depend only on the parameters of the electrical elements building the n-port, e.g. capacitances or resistances. The often used matrix is the Z-matrix, which handles again the voltages and currents instead of the a_i and b_i waves:

$$\begin{pmatrix} u_1 \\ u_2 \end{pmatrix} = Z \cdot \begin{pmatrix} i_1 \\ i_2 \end{pmatrix} = \begin{pmatrix} Z_{11} & Z_{12} \\ Z_{21} & Z_{22} \end{pmatrix} \begin{pmatrix} i_1 \\ i_2 \end{pmatrix}. \tag{2.2.32}$$

Both matrices are related by the following formula:

$$S = E - 2 \cdot (Z_N + E)^{-1}, \tag{2.2.33}$$

where E is the unit matrix and the matrix Z_N in the case of the 2-port is given by

$$Z_N = \begin{pmatrix} \frac{Z_{11}}{\sqrt{Z_1}\sqrt{Z_1}} & \frac{Z_{12}}{\sqrt{Z_1}\sqrt{Z_2}} \\ \frac{Z_{21}}{\sqrt{Z_2}\sqrt{Z_1}} & \frac{Z_{22}}{\sqrt{Z_2}\sqrt{Z_2}} \end{pmatrix}, \tag{2.2.34}$$

where Z_1 and Z_2 are the reference impedances.

With a network analyzer S-matrix of a 2-port can be easily measured. Since the second port of a VCSEL representing the active region is not achievable for a direct connection, only the S_{11} parameter can be investigated. That means that electrical parasitics of a VCSEL should be characterized based only on the measurement of electrical reflection of the incoming signal. That is why the 2-port describing the electrical parasitics of a VCSEL can not be determined from the incomplete S-matrix measurement, and some assumptions about the form of the parasitics should be met. In the most cases electrical parasitics of an oxide-confined VCSEL can be described by the equivalent circuit shown in Fig. 2.27b. This 2-port has three passive elements, two of which have the well known frequency dependence and R_m is frequency independent. The Z-matrix of this 2-port can be easily written and has the form

$$Z = \begin{pmatrix} \frac{R_1 \cdot (R_2 + R_3)}{R_1 + R_2 + R_3} & \frac{R_1 \cdot R_3}{R_1 + R_2 + R_3} \\ \frac{R_1 \cdot R_3}{R_1 + R_2 + R_3} & \frac{R_3 \cdot (R_1 + R_2)}{R_1 + R_2 + R_3} \end{pmatrix}, \tag{2.2.35}$$

with R_1, R_2 and R_3 defined as follows:

$$R_1 = -\frac{j}{2\pi C_p}, \tag{2.2.36}$$

$$R_2 = R_m, \tag{2.2.37}$$

$$R_3 = -\frac{j}{2\pi C_a}, \tag{2.2.38}$$

where j is the imaginary unit. Then using Eqs. 2.2.33 and 2.2.34 the S-matrix can be calculated setting Z_1 equal to 50 Ω and Z_2 equal to the resistance of the active region R_a. The calculated S_{11} can then be fitted to the measured real and imaginary parts of S_{11} and thus the four unknowns C_p, R_m, C_a and R_a can be estimated under the assumption that the used equivalent circuit represents electrical parasitics of the VCSEL correctly. Also more complicated n-ports can be simulated using advanced software, for example Microwave Office [43]. This gives the possibility to access individual parasitic elements directly and thus to analyze the electrical parasitics of VCSELs more efficiently. The frequency dependent S-parameter measurements up to 40 GHz were widely applied in the present work for VCSEL characterization and have helped to drastically improve the VCSEL performance.

2.2.5 Design of Impedance Matched High Frequency Contact Pads

After the VCSEL electrical parasitics have been considered in the previous section, in this section another important design issue will be described. In order to be able to guide the high frequency electrical driving signal to the VCSEL device, the laser should be equipped with appropriate contact pads, which should enable VCSEL operation at high frequencies. Since contact pads consist on high conductive metals, e.g. gold, chrome or platinum, the main problem to solve is not the attenuation of the signal but the impedance matching to the VCSEL driver or signal generator, which commonly have output standard impedances of 50 Ω. Thus contact pads should function at the same time as a high frequency transmission line. Free software for transmission line simulations is available, for example AppCAD [44] or TX Line [45], and in this work both AppCAD and TX Line were applied, delivering very similar results. In the following simulations carried out with TX Line for high speed contact pads designing will be presented.

There are different types of high speed transmission lines. Commonly, for VCSELs the so-called ground-signal-ground (GSG) transmission line is applied. Figure 2.29 shows schematically the GSG line with appropriate design parameters.

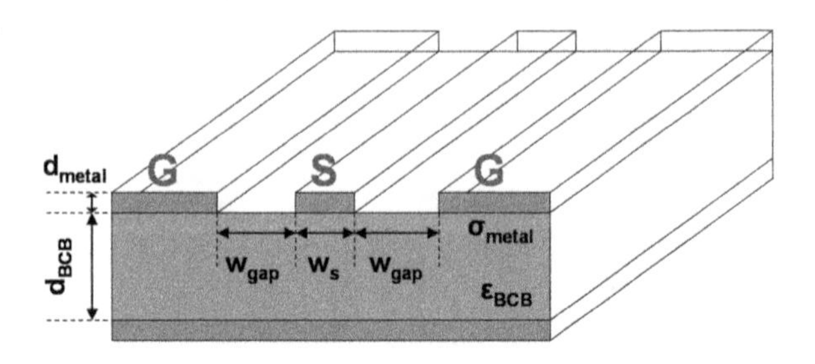

Fig. 2.29 Schematic view of a GSG transmission line with the corresponding parameters

A GSG transmission line consist of three metal lines: one for signal and two for the ground, which are placed symmetrically to the signal line and theoretically have infinite widths. The main design parameters are the width of the signal line w_s, the width of the gap between the signal and each ground line w_{gap}, the thickness of the dielectric, which was in our case bisbenzocyclobutene (BCB) d_{BCB} and the thickness of the applied metal d_{metal}. Also the dielectric constant of BCB ε_{BCB} and the metal conductivity σ_{metal} play an important role. All parameters were assumed to be frequency independent. Since all VCSELs investigated in this work were grown on doped semiconductor substrates, a GSG line having conductive material on the back side was simulated. Figure 2.30 shows simulation results for 5, 10 and 20 GHz with the corresponding parameters, which were used for VCSEL mask design for all VCSELs presented in this work.

From the figure one can see only weak frequency dependence of the transmission line impedance, which stays close to the goal impedance of 50 Ω received by using the parameters shown in the figure in red. Very important is the stability of the impedance to the variations of the design parameters, since these parameters can change during the fabrication process. Main reasons for this are limited precisions and reproducibility of the lithography steps but also of such important device fabrication techniques like metal deposition and BCB process. That is why the GSG-line should be designed in a way, that certain parameter changes would not drastically affect the transmission line impedance. In Fig. 2.31 simulation results for variations of all parameters and their effect on the transmission line impedance are presented.

From the figure one can see that the transmission impedance is very stable to design parameter changes. In a very wide variation range the impedance does not differ from the goal value of 50 Ω by more than 20%. For a better comparison the vertical impedance axis scaling is the same for all figures in Figs. 2.30 and 2.31.

After our investigations in this and also in the several previous sections we have seen, that electrical design of both epitaxial VCSEL structure, including the design of the active region, and also of the fabricated device is very important for a proper laser operation. Although, the physical processes, which should be considered

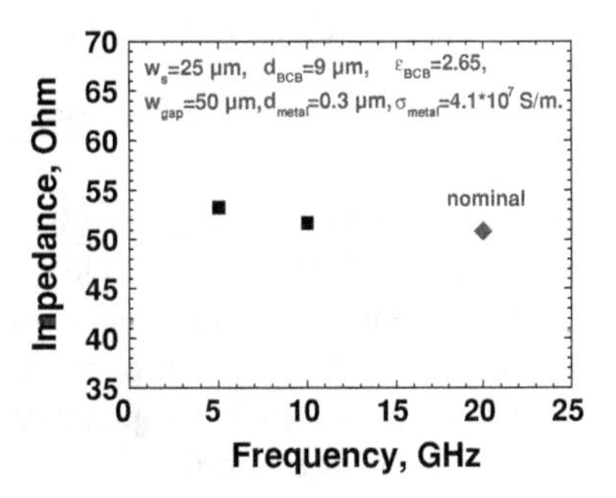

Fig. 2.30 Impedance for 5, 10 and 20 GHz with the corresponding nominal GSG-line parameters (*red*)

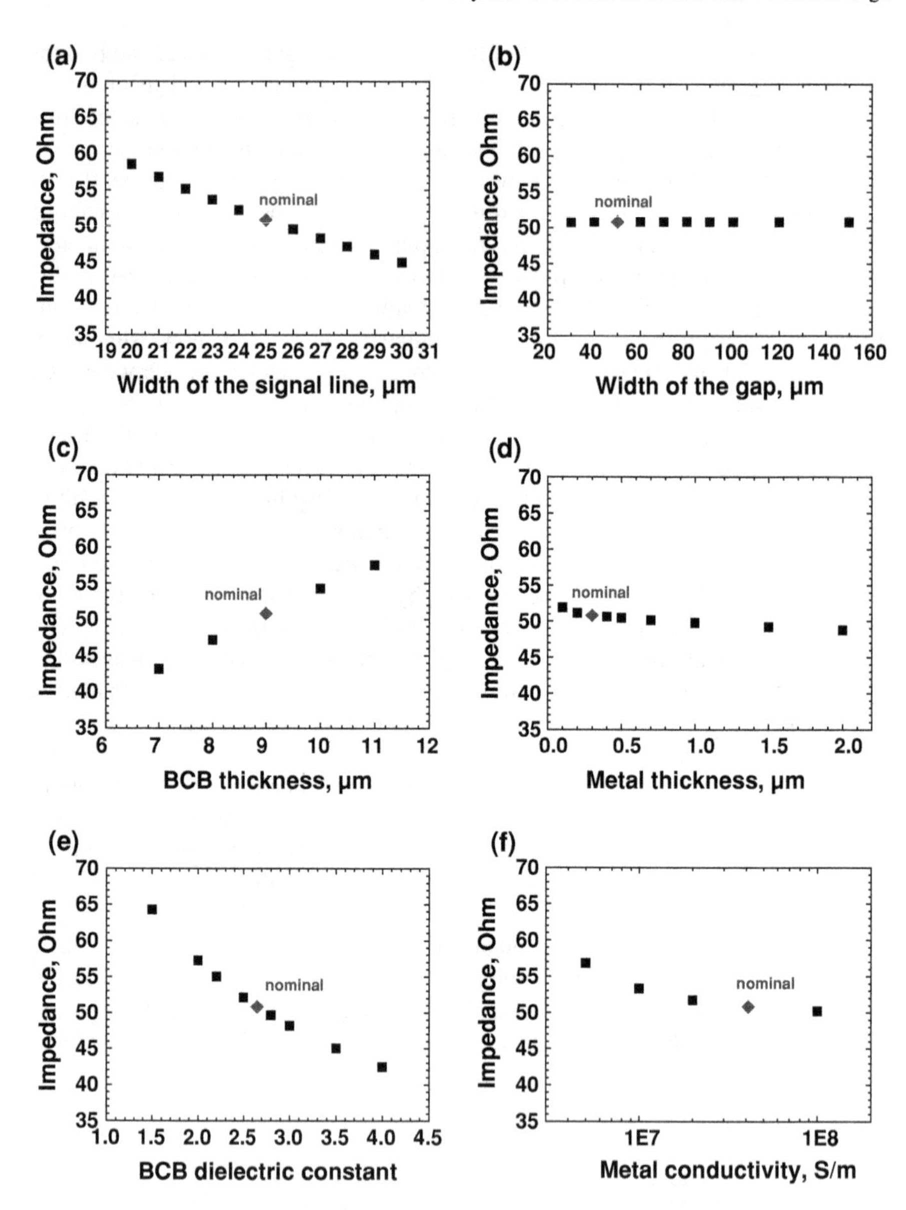

Fig. 2.31 Effects of the parameter variations on the transmission line impedance

during the VCSEL designing, are very complex, relatively simple methods could be applied in order to be able to solve all the design issues efficiently.

Optical and electrical phenomena build the basis of any semiconductor laser physics. Nevertheless each of them is affected in a very strongly manner by one physical quantity—the temperature. Temperature phenomena will be the main topic of the next section.

2.3 Thermal Properties

For a proper laser operation, not only at higher outer temperatures but also at room temperature, carefully designed thermal properties of the device are of the great importance, since the internal temperatures inside of the laser can be much higher than the ambience temperature. If semiconductor lasers operating at high outer temperatures, e.g. 85°C or higher, should be realized, thermal phenomena inside of the device become one of the crucial laser design aspects. It is of great importance to have a deeper understanding of the thermal processes in a laser, to know the heat generation and heat dissipation mechanisms. Because of the energy conservation, the part of the supplied electrical energy, which is not converted to optical energy inside of a laser, will generate heat in the device, strongly affecting device performance. Since efficiencies for converting electrical to optical power in common VCSELs are lower than 50%, the most part of the electrical energy introduced into the device will generate heat. The two major tasks of VCSEL thermal designing are to reduce the heat generation inside of the device and to carry off the heat, which have been generated, efficiently. Of course, it is better not to have the heat at all or to have only a small amount of it, than to generate much heat and to try to carry it off. In order to reduce heat generation, corresponding mechanisms should be studied. In a practical device a significant amount of the heat will be commonly still present, so heat flux phenomena should be considered as well. These mechanisms will be described in the present section.

2.3.1 Heat Generation and Thermal Resistance of Oxide-Confined VCSELs

Heat is generated in a semiconductor by transferring a part of the carrier energy to the lattice. These energy transfer processes are quantized and commonly described using quasi-particles called phonons, which then take part in heat dissipation processes. In the case of a local thermal equilibrium between lattice and carriers the heat flux is controlled by the temperature gradient [35]:

$$\vec{J}_{\mathrm{Heat}} = -\kappa_L \vec{\nabla} T, \tag{2.3.1}$$

where $\vec{J}_{\mathrm{Heat}}$ is the heat flux density (heat energy flux perpendicular to a unit surface), T is the local temperature and κ_L is the lattice thermal conductivity. In principle, electrons and holes contribute to the thermal conductivity, but their contribution is commonly small. The temperature should satisfy the heat flux equation

$$\rho_L C_L \frac{\partial T}{\partial t} = -\vec{\nabla} \cdot \vec{J}_{\mathrm{Heat}} + H_{\mathrm{Heat}}, \tag{2.3.2}$$

where ρ_L is the density of the semiconductor material, C_L is the lattice specific heat and $H_{\mathrm{Heat}}(\vec{r},t)$ is the generated heat power density from various sources. All parameters in 2.3.2 are in a common case temperature dependent by itself. If temperature changes are comparable small, these parameters can be approximately handled as corresponding constants. In principle, from the two Eqs. 2.3.1 and 2.3.2, by defined boundary conditions and heat generation function H_{Heat}, the two unknowns T and $\vec{J}_{\mathrm{Heat}}$ can be calculated.

In the static case the left hand side of (2.3.2) becomes zero and the heat flux is controlled only by one material constant, which is the thermal conductivity κ_L. The material of interest for the 850 and 980 nm VCSELs is $\mathrm{Al_xGa_{1-x}As}$, which is a ternary alloy. Because of the random distribution of alloy atoms, which causes strong allow scattering of phonons, the thermal conductivity of ternary and quaternary alloys is significant reduced as compared to pure binary materials like GaAs or AlAs. A good estimation of the thermal conductivity of an alloy of the form $\mathrm{AB_xC_{1-x}}$ can be found using the equation

$$\frac{1}{\kappa_L(x)} = \frac{x}{\kappa_{AB}} + \frac{1-x}{\kappa_{BC}} + x(1-x) \cdot C_{ABC}, \tag{2.3.3}$$

where κ_{AB} and κ_{BC} are thermal conductivities of the corresponding binary materials and C_{ABC} is the corresponding empirical bowing parameter. Figure 2.32 shows the thermal conductivity of $\mathrm{Al_xGa_{1-x}As}$ alloy calculated using Eq. 2.3.3 with corresponding material parameters [35] also shown in the picture.

As one can see the thermal conductivity goes down very fats with Al composition x and is already by 10% Al in AlGaAs reduced by more than a factor of two compared to pure binary GaAs. That is why using binary compounds are much more preferable from the point of view of the thermal properties of the lasers. Unfortunately at the wavelength of 850 nm pure GaAs is absorbing. That is why for the VCSELs emitting at this wavelength, like for the 850 nm QW-VCSELs

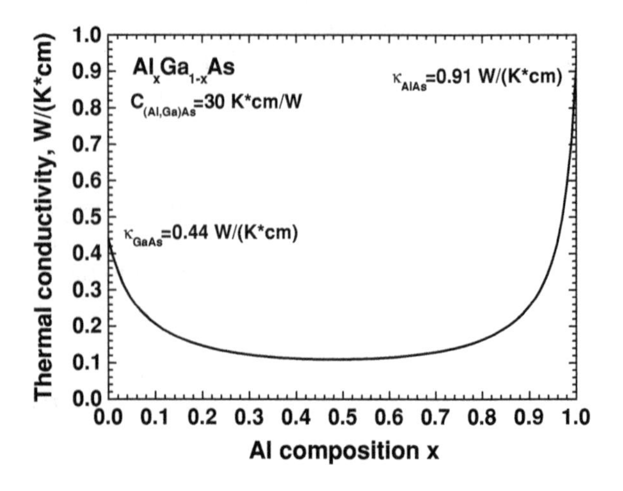

Fig. 2.32 Thermal conductivity of $\mathrm{Al_xGa_{1-x}As}$ as a function of the Al composition x, calculated using Eq. 2.3.3; corresponding parameters are also shown

presented in this work, some amount of Al should be used, for example 15%. For 980 nm pure GaAs can be applied, but there is another important limitation, which is the growth technique used for VCSEL growth. In the case of MOCVD binary GaAs is commonly grown at different temperatures as $Al_xGa_{1-x}As$ alloy, so that from the practical point of view it is easier to use also at the wavelength of 980 nm the ternary $Al_xGa_{1-x}As$ alloy. In the 980 nm QW-VCSELs investigated in this work 12% Al was used. By using MBE for the VCSEL structure growth binary GaAs can be easily applied. This was made for the 980 nm SML-VCSELs investigated in the present work.

From the other side, binary AlAs has an even higher thermal conductivity as the binary GaAs and would be preferable to use, especially in the top DBR mirror. The limiting factor here is in the case of oxide-confined VCSELs the oxidation rate of $Al_xGa_{1-x}As$, which is strongly dependent on the Al composition. Since the oxidation is carried out after the mesa etch, $Al_xGa_{1-x}As$ layers used in the DBR mirrors will also oxidize, and in order to keep the oxidation rate small lower Al composition as compared to the aperture layer should be used. Commonly 98–100% of Al is applied for the aperture layers, so that for the DBR layers the Al composition should be in the range of 90% or lower. In all VCSEL presented in this work 90% of Al in both DBR mirrors were applied.

To understand and effectively reduce the heat generation, corresponding phenomena should be considered. In a semiconductor laser heat is generated by different mechanisms. The most relevant are the Joule heat, the electron–hole recombination heat, the Thomson heat and the heat from optical absorption.

The Joule heat is generated by the carrier flow through a semiconductor and the corresponding scattering by phonons. The heat generation is then written by [35]

$$H_J = \frac{\vec{j}_n^{\,2}}{q\mu n} + \frac{\vec{j}_p^{\,2}}{q\mu p}, \tag{2.3.4}$$

which is directly proportional to the electrical resistance of the material.

The recombination heat is generated by nonradiative recombination processes of the electron–hole pairs. These processes include defect and impurity recombination, recombination on surfaces and interfaces and Auger recombination. The average recombination heat is proportional to the difference between the quasi-Fermi levels:

$$H_R = R\left(F_n - F_p\right), \tag{2.3.5}$$

where R is the total recombination rate of the considered processes. One could include also the spontaneous recombination, since most of the emitted photons are absorbed by the semiconductor and eventually converted into heat.

The Thomson heat results from the differences in the thermoelectric power, which is the measure for the increase in average carrier excess energy with increasing temperature. A dramatic example is the interface between different semiconductors, like in a DBR mirror. By moving through an interface from a material with the larger band gap to a material with the smaller band gap, electrons exhibit excess kinetic energy (hot electrons), which can dissipate to the lattice. In the opposite case electrons needs to receive some energy from the lattice. Thus the Thomson heat can be both positive and negative and is described by

$$H_T = -qT\left(\vec{j}_n \cdot \vec{\nabla} P_n + \vec{j}_p \cdot \vec{\nabla} P_p\right), \tag{2.3.6}$$

where P_n and P_p are the thermoelectric power for electrons and holes, respectively. Sometimes Thompson heat is also referred to as Peltier heat.

The optical absorption heat arises from the absorption of photons in the semiconductor material. Photons can be directly absorbed by the crystal lattice, but at the typical photon energies considered here the free carrier absorption is dominating. The absorbed energy is then quickly dissipated to the lattice due to very short intraband scattering times. The absorption heat is given by

$$H_A = \alpha \Phi h v, \tag{2.3.7}$$

where $\alpha(hv)$ is the wavelength dependent absorption coefficient, v is the photon frequency and Φ is the photon flux energy, which can be calculated from the optical simulations considered above.

By setting together all heat generation mechanisms described by Eqs. 2.3.4–2.3.7 one can calculate the total heat generation rate H_{Heat} and use it for simulations of the thermal properties according to Eqs. 2.3.1 and 2.3.2.

An important quantity characterizing thermal properties of a VCSEL is its thermal resistance. If the heat power is generated mostly within one region, e.g. the active region in a VCSEL, the temperature increase in this region will be determined by the thermal resistance of the whole device and the generated heat power

$$\Delta T = R_{\text{th}} P_{\text{Heat}}, \tag{2.3.8}$$

where ΔT is the temperature difference between the heat source (commonly active region) and the heat sink, P_{Heat} is the total heat power and R_{th} is the thermal resistance of the laser. The heat power is equal to the supplied electrical power minus the output optical power:

$$P_{\text{Heat}} = UI - P_{\text{out}}, \tag{2.3.9}$$

where U and I are the operating laser voltage and current and P_{out} is the total optical output power. The thermal resistance of a VCSEL can be easily estimated by measuring the shift of the wavelength first with the current, additionally to the L-U-I-characteristics, at one temperature and then near the threshold for several higher temperatures. From the first measurement set the wavelength shift in dependence on the heat power can be calculated, while form the second measurement set the wavelength shift as a function of the temperature is directly measured. Combining the results of both sets, temperature increase as a function of the dissipated heat power and then according to (2.3.8) the thermal resistance can be calculated. Also the temperature of the active region can be directly measured with this method. This enables to study and optimize high temperature laser performance more efficiently.

If one assumes the model of a uniform temperature disc, representing the active region of an oxide-confined VCSEL, on a homogeneous, isotropic, semi-infinite substrate, the thermal resistance can be given by the simple formula [46]:

$$R_{\text{th}} = \frac{1}{2\xi d}, \tag{2.3.10}$$

where ξ is the substrate thermal conductivity and d is the disc diameter. For a VCSEL, d corresponds to the effective pumped active region diameter, which is close to the aperture diameter. For ξ some effective value representing the bottom part of the cavity and the bottom DBR mirror should be used. If one assumes that the main heat source in the active region is the Joule heat, one can see from (2.3.4), that the heat power generated in the active region is directly proportional to the squared current flowing through it and the corresponding electrical resistance, mainly determined by the oxide aperture diameter d:

$$P_{\text{Heat}} = I_a^2 R_a = j_a^2 A^2 \rho_a \frac{l_a}{A} = j_a^2 \rho_a l_a \frac{\pi d^2}{4}, \tag{2.3.11}$$

where I_a is the current flowing through the active region, which in absence of large leakage currents is practically equal to the laser operation current I, j_a is the effective current density in the active region, A is the active region surface, R_a is the electrical resistance of the active region, ρ_a is the effective characteristic electrical resistance of the active region material and l_a is the active region effective length. If one then combines (2.3.8) and (2.3.10) using (2.3.11) one gets the following result for the temperature in the active region as a function of the aperture diameter:

$$\Delta T = \frac{1}{2\xi d} j_a^2 \rho_a l_a \frac{\pi d^2}{4} = C_{\text{eff}} j_a^2 d, \tag{2.3.12}$$

where C_{eff} represents all the constants describing the VCSEL structure and effectively independent on the current and aperture diameter:

$$C_{\text{eff}} = \frac{\rho_a l_a \pi}{8\xi}. \tag{2.3.13}$$

From Eq. 2.3.13 follows that for a constant current density in the active region j_a under described assumptions the temperature increase in VCSELs with larger aperture diameters d is larger as compared to VCSELs with smaller aperture diameters. This means that VCSELs with smaller apertures can operate at higher current densities at the same temperature as VCSELs with larger apertures.

2.3.2 *Temperature Dependence of the Basic Laser Parameters*

Since temperature affects practically each physical process in a semiconductor laser, it is inalienable to understand temperature dependence of the basic laser parameters. We will see in this section, that temperature effects have a very strong impact on laser operation not only at higher temperatures but also at room temperature, because, as we have seen in the previous section, the largest part of the introduced energy is converted to heat inside of a laser. This can be seen already for CW characteristics measured at room temperature, like L-I-curve and optical emission spectra. Figure 2.33 shows temperature dependent L-U-I characteristics and extracted values for the threshold current, maximum output power and maximum differential efficiency for temperatures from 20 to 100°C for an 850 nm QW-VCSEL.

One can see that device performance is strongly temperature dependent. The threshold current increases by more than twice for temperature changes between 20 and 100°C, the maximum output power decreases similarly and the differential efficiency drops also by more than 40%. But also at 20°C the output power does not stay a linear function, as predicted by the rate equation theory without temperature effects [1, 2]. The output power saturates at some point, called thermal roll-over, and then starts to decrease.

In order to understand temperature effects on laser operation qualitatively we should take a closer look on the laser equations. One of the most important characteristics of a semiconductor laser is its L-I-curve, this means its output power as a function of the driving current [1, 2]:

$$P_0 = \eta_d \frac{h\nu}{q}(I - I_{\text{th}}), \tag{2.3.14}$$

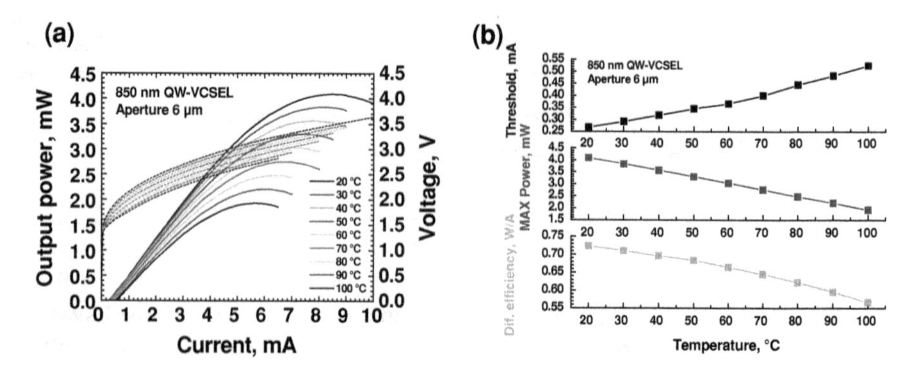

Fig. 2.33 Temperature dependent L-U-I characteristics (**a**) and extracted values for the threshold current, maximum output power and maximum differential efficiency (**b**) from 20 to 100°C for an 850 nm QW-VCSEL

where P_0 is the laser output power, η_d is the differential efficiency I is the driving current and I_{th} is the threshold current. The differential efficiency η_d can be expressed as follows:

$$\eta_d = \frac{\eta_i \alpha_m}{\alpha_i + \alpha_m}, \tag{2.3.15}$$

where η_i is the internal quantum efficiency or injection efficiency, which is the fraction of the terminal current that generates carrier in the active region, α_i is the internal or intrinsic cavity loss and α_m is the mirror loss, both already introduced in Eq. 2.1.23.

According to Eq. 2.3.14 the output power should be a linear function of the injected current. However, as one can see from Fig. 2.33, in the reality this is not the case even for one constant outer temperature. The reason is the internal heating, which increases the temperature inside of the device. The internal losses increase, approximately following the linear relationship $\alpha_i \propto T$ [2], which leads to a decrease of the differential efficiency according to (2.3.15). Exactly this experimentally measured behavior of the differential efficiency is shown in Fig. 2.33. Because the peak material gain for a given carrier density decreases, higher carrier densities are required to compensate increased losses and to achieve lasing. Accordingly the threshold current increases, which can be also seen in the Fig. 2.33.

The differential resistance of the laser commonly decreases with increasing temperature, because the conductivity of the semiconductor materials applied for laser fabrication increases. This can be seen also in the Fig. 2.33a, where the voltage at some constant defined current decreases with increasing temperature.

Finally the output power saturates at some internal temperature, because the incoming current can not compensate the increase of the threshold current and the drop of the differential efficiency any more. For VCSELs, important is the dependence of the roll-over current I_{ro} on the aperture diameter. If we assume that for any VCSEL from the same piece the thermal roll-over occurs at one defined temperature T_{ro}, we can easily derive the following dependence for a given outer or chuck temperature T_{chuck} using (2.3.8) and (2.3.10):

$$\Delta T_{\mathrm{ro}} = T_{\mathrm{ro}} - T_{\mathrm{chuck}} = R_{\mathrm{th}} I_{\mathrm{ro}}^2 R_a = \frac{2\rho_a l_a}{\xi\pi} \frac{I_{\mathrm{ro}}^2}{d^3}, \tag{2.3.16}$$

with already used in (2.3.11) electrical resistance of the active region R_a at the roll-over temperature, effective characteristic electrical resistance of the active region material ρ_a at the roll-over temperature and active region effective length l_a. Hereby d is the aperture diameter, if no current crowding is assumed, and ξ is the substrate thermal conductivity. We can rewrite Eq. 2.3.16 as follows:

$$I_{\mathrm{ro}}^{2/3} = \sqrt{[3]} \frac{\xi\pi}{2\rho_a l_a} d = C_{\mathrm{ro}} d. \tag{2.3.17}$$

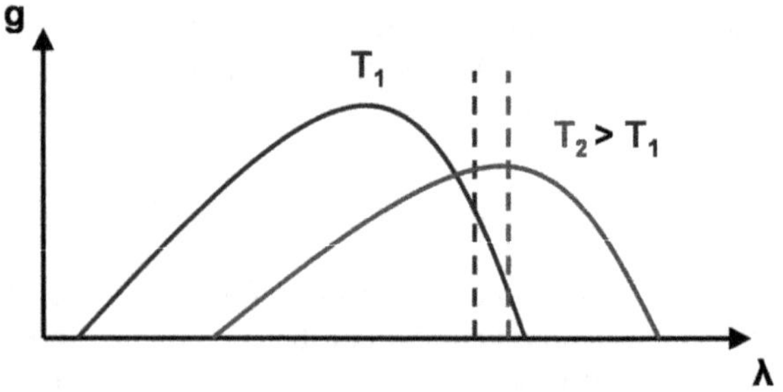

Fig. 2.34 Schematic illustration of changes in the cavity-gain detuning with increasing temperature; dashed lines show the wavelength of the cavity dip at both temperatures

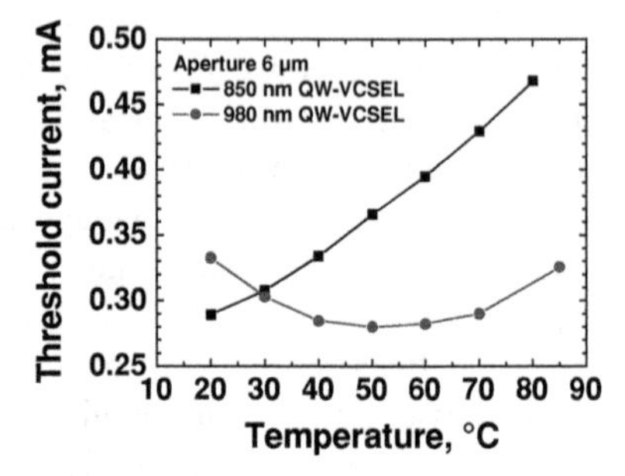

Fig. 2.35 Threshold current as function of temperature for a 850 nm QW-VCSEL and a 980 nm QW-VCSEL both with 6 μm aperture diameter

As one can see, the dependence of the roll-over current on the aperture diameter at a given outer temperature is not linear, but as one can see both experimentally and from Eq. 2.3.17 for larger apertures the roll-over current is also larger.

For VCSELs, where the lasing wavelength is defined not by the gain peak wavelength but by the cavity dip position, an additional important effect should be considered. This is the detuning between the cavity mode position and the peak gain. Because these both values have not identical temperature dependences, their difference changes with temperature. In Fig. 2.34 schematic illustration of this process is shown. As temperature increases, the gain peak commonly drops and shifts faster than the cavity dip, shown as dashed lines, so that the cavity-dip detuning decreases. It can happen that at a higher temperature more gain is available at the cavity wavelength, like in the schematic example shown in Fig. 2.34.

Knowing this one can design a VCSEL with the cavity-gain detuning optimized for a specific temperature. Figure 2.35 shows an example of such optimization.

In the figure threshold current for two VCSELs with nominally identical aperture diameter of 6 μm is shown. In the case of 850 nm QW-VCSEL the cavity-gain detuning was optimized for room temperature operation, while in the case of 980 nm QW-VCSEL optimal temperature was designed to be around 50°C. One can see, that the threshold current of the first VCSEL increases continuously with temperature, while for the 980 nm QW-VCSEL the threshold current decreases for temperatures up to 50°C, and then starts to increase, reaching at 85°C the value similar to 20°C. The overall changes are ~62% for the 850 nm

QW-VCSEL optimized for room temperature and only ~19% for the 980 nm QW-VCSEL optimized for high temperature operation.

Temperature effects shortly described in this section play a major role and should be carefully considered during VCSEL designing process. Of course temperature affects not only CW characteristics of a laser. Also high frequency properties are affected in a very strongly manner. As we will see in further sections, one can efficiently design and fabricate high speed VCSELs operating at very high temperatures.

2.4 The Rate Equations

After optical, electrical and thermal properties of oxide-confined VCSELs were briefly investigated, in this section the temperature independent single-mode rate equation model will be described. This is a simple, but very powerful model, which describes dynamics of charge carriers and photons in a semiconductor laser [1, 2]. For the multimode case it has been demonstrated, both theoretically and experimentally, that index guided multimode VCSELs with highly overlapping transverse fields (such as oxide-confined VCSELs) have uniform carrier and photon densities and exhibit a single resonance frequency [47, 48], so that the single-mode rate equation model could be applied [49]. Because of its evidence and clearness the model is very helpful for understanding of the static and dynamic behavior of semiconductor lasers as well as for numerical qualitative and quantitative simulations of laser devices. By fitting model results to experimental data one gets access to important physical quantities describing a VCSEL, e.g. differential gain, relaxation resonance frequency, damping etc. Application of the rate equation model for analyzing real VCSELs is today a standard procedure for device understanding and improvement.

2.4.1 Rate Equation Model and Steady-State Solutions

In the rate equation model two types of particles play a role: charge carriers and photons. Because in mostly cases active region of a VCSEL is undoped or only lightly doped, under high injection levels, relevant for lasers, the charge neutrality requires the density of negative carriers (electrons) to be equal to the density of positive carriers (holes). In such cases one can investigate the density of only one type of carriers, for example of the electrons. This fact greatly simplifies the analysis.

In a semiconductor lasers there are several physical processes affecting dynamics of the carriers and photons. Figure 2.36 shows these processes used in the presented rate equation model. We will describe carriers and photons by their densities (expressed e.g. in units of $1/cm^3$): N for electrons and N_p for photons.

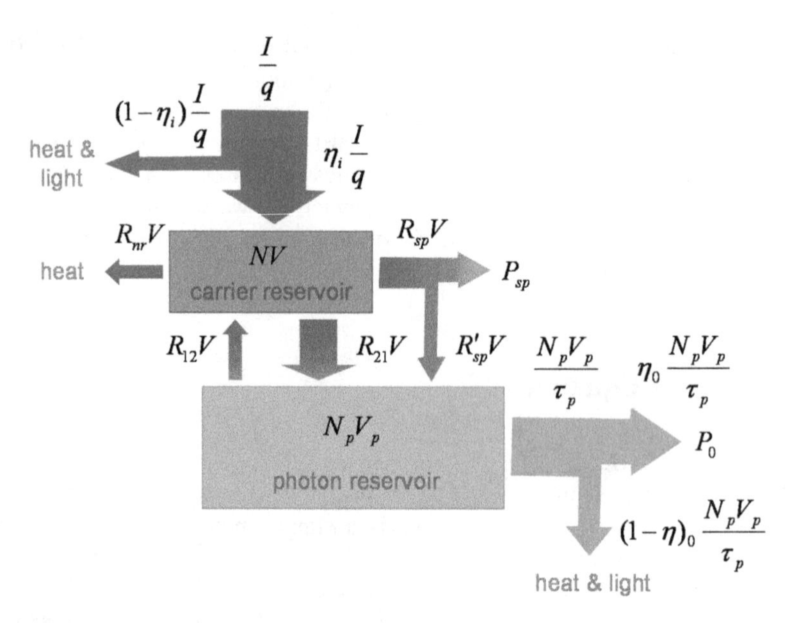

Fig. 2.36 Carrier and photon generation, recombination and transformation processes used in the rate equation model. *Blue arrows* correspond to carriers and *orange arrows* correspond to photons

The active region is hereby the region, where recombining carriers contribute to useful gain and photon emission, and corresponds in Fig. 2.36 to the carrier reservoir with the volume V. The photon reservoir is the lasing mode with an effective volume V_p, which is identical to the volume V_{eff} used in Eq. 2.1.44, and can be calculated as described in the previous sections. We will focus here on single-mode devices, but the rate equation theory can be easily adopted for the multimode case. In Fig. 2.36 for a better understanding the number of particles is shown, so that arrows describe the particle flow (particles per unit time, in units of e.g. 1/s), and boxes corresponds to the number of particles (which is dimensionless). If one summarizes all the processes shown in the figure, one arrives at the carrier and photon number rate equations:

$$V\frac{dN}{dt}=\frac{\eta_i I}{q}-\left(R_{\mathrm{sp}}+R_{\mathrm{nr}}\right)V-(R_{21}-R_{12})V, \tag{2.4.1}$$

$$V_p\frac{dN_p}{dt}=(R_{21}-R_{12})V-\frac{N_pV_p}{\tau_p}+R'_{\mathrm{sp}}V. \tag{2.4.2}$$

Each term in (2.4.1) and (2.4.2) can be easily found in the Fig. 2.36. The number of carriers increases because of the first term on the right hand side in (2.4.1), which corresponds to the carrier injection by current. Here η_i is the internal quantum efficiency or injection efficiency, already introduced in (2.3.15), which is the fraction of the terminal current that generates carriers in the active region, and q is the magnitude of the electron charge. Spontaneous and non-radiative recombination described by the spontaneous and non-radiative recombination rates R_{sp} and R_{nr} per unit volume and time, respectively, which have units of 1/(cm^3 s), are the major parasitic carrier loss mechanisms, that do not contribute noticeable to the photon density in the lasing mode. Finally, the last term in (2.4.1) describes the carrier loss (or generation) by the stimulated emission (or absorption) process. Hereby R_{21} is the stimulated emission rate per unite volume for photon emission into the lasing mode, and R_{12} is the stimulated absorption rate per unit volume for the photons from the lasing mode.

The last term of the (2.4.1) appears again on the right hand side of the (2.4.2), but this time it corresponds not to the loss of the carriers but to the increase of the photon number in the lasing mode. Another term describing the photon increase is the last term on the right hand side of (2.4.2), which corresponds to the part of photons generated by spontaneous emission, which goes into the lasing mode. Thereby R'_{sp} is the rate per unit volume of spontaneous emission into the lasing mode and can be described by the following equation:

$$R'_{sp} = \beta_{sp} R_{sp}, \tag{2.4.3}$$

where β_{sp} is the fraction of the photons generated by spontaneous emission, which goes into the lasing mode. In common VCSELs β_{sp} is in the order of 10^{-4}. By applying special cavity designs this value can be increased, leading to more efficient spontaneous recombination processes. Finally, the only term which corresponds to the photon losses is the second term on the right hand side of (2.4.2), which is dominated by the photon lifetime τ_p. The photon lifetime τ_p is given by

$$\frac{1}{\tau_p} = v_g(\alpha_m + \alpha_i) = \frac{\omega}{Q}, \tag{2.4.4}$$

where v_g is the group velocity of the mode of interest, including both material and waveguide dispersion, and α_m and α_i are the mirror and the spatial averaged internal cavity losses, already introduced in (2.1.23). Photon lifetime can be defined also in terms of the light frequency ω and the cavity quality factor Q, which sometimes can be advantageous. The mirror loss can be calculated using the optical properties of the cavity according to

$$\alpha_m = \frac{1}{2L} \ln\left(\frac{1}{R_1 R_2}\right), \tag{2.4.5}$$

where L is the effective cavity length and R_1 and R_2 are the power reflectivities of the top and bottom mirror in the case of VCSELs.

All recombination rates in (2.4.1) and (2.4.2) are dependent on the carrier density N. The non-radiative recombination rate R_{nr} can be expressed by

$$R_{nr} = AN + CN^3, \tag{2.4.6}$$

where constant coefficients A and C corresponds to different non-radiative recombination mechanisms, e.g. defect and impurity recombination, surface and interface recombination and Auger recombination. In mostly VCSELs A could be assumed to be negligible. The constant C, describing the Auger recombination, is in the order of 10^{-29}–10^{-30} cm^6/s in the active regions used in common GaAs- and InGaAs-based VCSELs. The spontaneous emission rate R_{sp} has the following dependence on carrier density:

$$R_{\mathrm{sp}} = BN^2, \tag{2.4.7}$$

where B is the bimolecular recombination coefficient and has a magnitude of $\sim 10^{10}$ cm^3/s for the most AlGaAs and InGaAsP alloys of interest.

The last term in (2.4.1), which is the first term on the right hand side of the (2.4.2), is the photon gain term and can be expressed using the following equation:

$$g = \frac{1}{N_p}\frac{dN_p}{dz} = \frac{1}{v_g N_p}\frac{dN_p}{dt} = \frac{1}{v_g N_p}(R_{21} - R_{12}), \tag{2.4.8}$$

where g is the material gain describing increase (or decrease, if negative) of the photon density N_p by passing through the active region. According to (2.4.8), the photon gain term in (2.4.1) and (2.4.2) can be expressed by

$$(R_{21} - R_{12}) = v_g g N_p. \tag{2.4.9}$$

The gain g is in a common case a function of both carrier and photon densities and can be approximated at a given wavelength by a logarithmic function

$$g(N, N_p) = \frac{g_0}{1 + \varepsilon N_p}\ln\left(\frac{N + N_s}{N_{tr} + N_s}\right), \tag{2.4.10}$$

where g_0, N_s and N_{tr} are constants and ε is the empiric gain compression factor, which is a constant as well.

One can rewrite (2.4.1) and (2.4.2) by dividing out the corresponding volumes and using the following equation:

$$\Gamma = \frac{V}{V_p}, \tag{2.4.11}$$

where Γ is the optical confinement factor already introduces in (2.1.23) and (2.1.45), and also using (2.4.9), to obtain the density rate equations, which are more familiar:

$$\frac{dN}{dt} = \frac{\eta_i I}{qV} - (R_{\mathrm{sp}} + R_{\mathrm{nr}}) - v_g g N_p, \tag{2.4.12}$$

$$\frac{dN_p}{dt} = \left(\Gamma v_g g - \frac{1}{\tau_p}\right)N_p + \Gamma R'_{\mathrm{sp}}. \tag{2.4.13}$$

Putting (2.4.6) and (2.4.7) together and making the same with (2.4.4) and (2.4.5) we get

$$R_{\mathrm{sp}} + R_{\mathrm{nr}} = AN + BN^2 + CN^3, \tag{2.4.14}$$

$$\frac{1}{\tau_p} = v_g\left(\frac{1}{2L}\ln\left(\frac{1}{R_1R_2}\right) + \alpha_i\right). \tag{2.4.15}$$

Equations (2.4.12–2.4.15) together with (2.4.3) combined with (2.4.7), leading to Eq. 2.4.17, and also Eq. 2.4.10 which we will repeat here for more clearness as (2.4.16),

$$g(N, N_p) = \frac{g_0}{1 + \varepsilon N_p}\ln\left(\frac{N + N_s}{N_{tr} + N_s}\right), \tag{2.4.16}$$

$$R'_{\mathrm{sp}} = \beta_{\mathrm{sp}}R_{\mathrm{sp}} = \beta_{\mathrm{sp}}BN^2, \tag{2.4.17}$$

build the equation set (2.4.12–2.4.17) for the carrier density N and the photon density N_p, where all coefficients, excepted for the driving current I, are constants or can be expressed as functions of N and N_p. By solving this set of equations (2.4.12–2.4.17) one can investigate laser behavior under different types of current injection: constant current, small signal modulation, large signal modulation etc.

To get access to the output power of the lasing mode of interest, one can use the following equations:

$$P_0 = \eta_0 h\nu\frac{N_pV_p}{\tau_p}, \tag{2.4.18}$$

$$\eta_0 = F\frac{\alpha_m}{\alpha_m + \alpha_i}, \tag{2.4.19}$$

where η_0 is the optical efficiency, $h\nu$ is the photon energy and F is the fraction of power not reflected back into the cavity which escapes as useful power from the output coupling mirror. By multiplying the optical efficiency η_0 with the injection efficiency η_i one gets the differential efficiency η_d, defined earlier in (2.3.15), which can be easily measured experimentally:

$$\eta_d = \eta_i\eta_0. \tag{2.4.20}$$

The factor F is for common VCSELs close to 1, that is why it was suppressed in (2.3.15).

By setting the time derivatives for the carrier and photon density equal to zero in the equation set (2.4.12–2.4.17), one gets the steady-state equations describing laser behavior under constant current injection. The steady-state solution is

$$N_p = \frac{\Gamma R'_{\mathrm{sp}}}{1/\tau_p - \Gamma v_g g}, \tag{2.4.21}$$

$$I = \frac{qV}{\eta_i}\left(R_{\mathrm{sp}} + R_{\mathrm{nr}} + v_g g N_p\right). \tag{2.4.22}$$

Here one can use the carrier density N as the independent variable for more physical understanding. One can apply following equations

$$\Gamma v_g g_{\text{th}} \equiv \frac{1}{\tau_p}, \tag{2.4.23}$$

$$g(N_{\text{th}}) = g_{\text{th}}, \tag{2.4.24}$$

to define the threshold gain g_{th} and the corresponding threshold carrier density N_{th}. With this definitions one can solve for $v_g g N_p$ in (2.4.21) and by setting the result to (2.4.22) one gets the following expression for the current:

$$I = \frac{qV}{\eta_i}\left((1-\beta_{\text{sp}})R_{\text{sp}} + R_{\text{nr}} + v_g g_{\text{th}} N_p\right). \tag{2.4.25}$$

For the case when the laser is driven well above threshold, one can use threshold values by applying $N \rightarrow N_{\text{th}}$ for all parameters in (2.4.21) and (2.4.25) except for the g in the denominator of (2.4.21), leading to

$$N_p(N) = \frac{R'_{\text{sp}}(N_{\text{th}})/v_g}{g_{\text{th}} - g(N)}, \tag{2.4.26}$$

$$I = \frac{qV}{\eta_i}\left((1-\beta_{\text{sp}}(N_{\text{th}}))R_{\text{sp}}(N_{\text{th}}) + R_{\text{nr}}(N_{\text{th}})\right) + \frac{qV}{\eta_i} v_g g_{\text{th}} N_p(N). \tag{2.4.27}$$

It is clear from (2.4.26) that the carrier density N and gain g actually never reach their threshold values N_{th} and g_{th} for finite output powers and currents. They remain ever below these values.

From (2.4.18) one can rewrite the output power using the definition (2.4.23) and (2.4.11) as follows:

$$P_0 = \eta_0 h\nu\left(V v_g g_{\text{th}} N_p\right). \tag{2.4.28}$$

By solving for $V v_g g_{\text{th}} N_p$ in (2.4.28) and putting the result into the last term of (2.4.27) one can get after some rearrangement the output power in the lasing mode P_0 as a function of the driving current I, the so-called L-I-curve:

$$P_0 = \eta_i \eta_0 \frac{h\nu}{q}(I - I_{\text{th}}), \tag{2.4.29}$$

where the threshold current I_{th} is defined by

$$I_{\text{th}} = \frac{qV}{\eta_i}\left((1-\beta_{\text{sp}}(N_{\text{th}}))R_{\text{sp}}(N_{\text{th}}) + R_{\text{nr}}(N_{\text{th}})\right). \tag{2.4.30}$$

One can see from (2.4.29) that the output power in the lasing mode increases linearly with the injected current above threshold. The threshold current is determined according to (2.4.30) by the spontaneous and non-radiative recombination rates at the threshold carrier density N_{th}, which is defined by the mirror and internal losses corresponding to (2.4.23) and (2.4.15) and by the form of the gain dependence on the carrier and photon densities (2.4.16). If the non-radiative recombination rate R_{nr} is negligible and $\beta_{sp} = 1$ the threshold current would be equal to zero and the so-called "thresholdless" laser would be realized. However, for typical lasers $\beta_{sp} \ll 1$ and the threshold current remains finite.

To illustrate the behavior of the carrier density, photon density, gain and output power during the laser steady-state operation, their dependence on the injected current is shown in Fig. 2.37.

As one can see from the figure, the photon density N_p and accordingly the output power in the lasing mode show a drastic increase as the injected current reaches its threshold value (denoted by the vertical line near 400 μA). Above threshold the photon density and the output power increase linearly with the current, as predicted by (2.4.27) and (2.4.29). The carrier density N and the gain g nearly saturate at threshold and increase at currents above threshold only by a small amount.

As one example of application of the rate equations one can take a look on the simulated L-I-curves for different values of the reflectivity of the top mirror of an oxide-confined VCSEL and also for different internal cavity losses, shown in Fig. 2.38. The first three L-I-curves correspond to the same value of internal losses α_i equal to 20 cm^{-1}, but to different top mirror reflectivities of 99.7, 99.6 and 99.5%. Accordingly to (2.4.23) and (2.4.15) the threshold gain increases with increased mirror losses (lower top mirror reflectivities) leading to an increase in the threshold current. The differential efficiency η_d, represented by the slope of the L-I-curves, increases as well, as predicted by (2.4.19) and (2.4.20).

A different situation exists if not the mirror losses α_m but the internal losses α_i are increased, as shown by the fourth L-I-curve in Fig. 2.38. Here again the threshold current increases, but the differential efficiency decreases, as can be easily seen from (2.4.19) and (2.4.20). Thus by investigating measured L-I-curves and applying results of the rate equation theory one can distinguish between different physical processes in a semiconductor laser. Numerical simulations

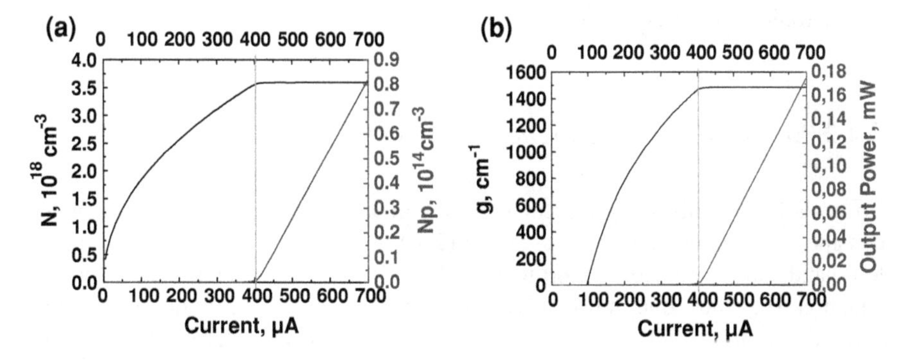

Fig. 2.37 Simulated carrier density N and photon density N_p (**a**) and gain g and output power (**b**) as a function of the injected current I; the *vertical line* denotes the position of the threshold current

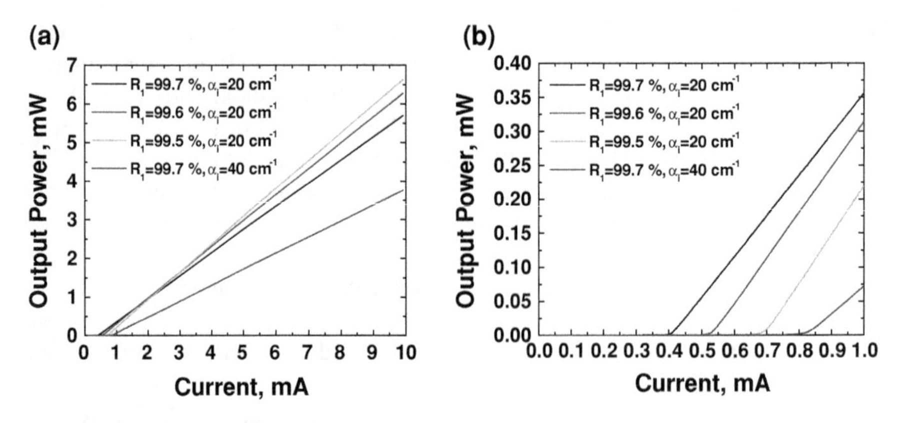

Fig. 2.38 Simulated L-I characteristics for VCSELs with different reflectivities of the top mirror and different internal cavity losses (**a**) and the zoomed view (**b**)

implementing rate equations are very useful for understanding of the physical processes inside of a semiconductor laser. Combined with a deeper analysis of the corresponding equations they provide a very powerful framework for successful VCSEL designing.

In this section rate equations were introduced without paying attention to temperature effects, that is why the simulated L-I-curves, e.g. those shown in Fig. 2.38, are strictly linear and do not behave similar to measured L-I-curves, e.g. those shown in Fig. 2.33. In principle one can add temperature effects by introducing temperature dependence for corresponding physical parameters.

Understanding of the principles of the laser operation under the constant current injection, i.e. in the CW regime, is an indispensable precondition for successful VCSEL development. Additionally, for high-speed VCSELs, like for those investigated in this work, modulation properties at high frequencies are the major issue to be understood and improved. That is why the next important application of the rate equations is the simulation of the semiconductor laser dynamics. This will be the topic of the next section.

2.4.2 Small-Signal Frequency Response

To be able to analyze dynamical behavior of semiconductor lasers one should consider the rate equations (2.4.12) and (2.4.13) with the time derivatives on the left hand side included. Unfortunately there is no analytical solution for this equation set, thus one must solve the equations numerically or apply some approximations to get analytical solutions. One of such approximations is the assumption that the changes in the carrier and photon densities away from their steady-state values are small compared to these steady-state densities. Then one can investigate the so-called small-signal responses of one variable in terms of perturbation to another and get insight into the dynamic processes in the laser. For this purpose one can take differential of both (2.4.12) and (2.4.13), considering the injection current I, the carrier density N, the photon density N_p and the gain g as dynamic variables:

$$d\left[\frac{dN}{dt}\right] = \frac{\eta_i}{qV}dI - \frac{1}{\tau_{\Delta N}}dN - v_g g dN_p - N_p v_g dg, \tag{2.4.31}$$

$$d\left[\frac{dN_p}{dt}\right] = \left(\Gamma v_g g - \frac{1}{\tau_p}\right)dN_p + N_p \Gamma v_g dg + \frac{\Gamma}{\tau'_{\Delta N}}dN. \tag{2.4.32}$$

Hereby the differential carrier lifetime $\tau_{\Delta N}$ and the differential lifetime of carriers, which radiate photons into the lasing mode, $\tau'_{\Delta N}$ were introduced according to the following equations:

$$\frac{1}{\tau_{\Delta N}} = \frac{dR_{\mathrm{sp}}}{dN} + \frac{dR_{\mathrm{nr}}}{dN} = A + 2BN + 3CN^2, \tag{2.4.33}$$

$$\frac{1}{\tau'_{\Delta N}} = \frac{dR'_{\mathrm{sp}}}{dN} = 2\beta_{\mathrm{sp}}BN + \frac{d\beta_{\mathrm{sp}}}{dN}BN^2. \tag{2.4.34}$$

The differential carrier lifetime for the carriers emitting into the lasing mode $\tau'_{\Delta N}$ is typically in the range of tens of microseconds, and its contribution is negligible in the most cases.

One can expand the gain differential, paying attention to the fact that gain depends both on carrier and photon densities:

$$dg = adN - a_p dN_p, \tag{2.4.35}$$

where the gain derivative regarding to the carrier density a and the negative gain derivative regarding to the photon density a_p were introduced. If one assumes the gain dependence like in (2.4.16), following equations for the gain derivatives will hold:

$$a = \frac{\partial g}{\partial N} = \frac{g_0}{(N + N_s)(1 + \varepsilon N_p)}, \tag{2.4.36}$$

$$a_p = -\frac{\partial g}{\partial N_p} = \frac{\varepsilon g}{(1 + \varepsilon N_p)}, \tag{2.4.37}$$

where both a and a_p are not constant but depend on carrier and photon densities. Inserting (2.4.35) into (2.4.31) and (2.4.32) and rearranging several terms one gets following equations, where only dI, dN and dN_p are present as independent differentials:

$$d\left[\frac{dN}{dt}\right] = \frac{\eta_i}{qV}dI - \left(\frac{1}{\tau_{\Delta N}} + N_p v_g a\right)dN - \left(v_g g - N_p v_g a_p\right)dN_p, \tag{2.4.38}$$

$$d\left[\frac{dN_p}{dt}\right] = \left(\frac{\Gamma}{\tau'_{\Delta N}} + N_p\Gamma v_g a\right)dN - \left(\frac{1}{\tau_p} - \Gamma v_g g + N_p\Gamma v_g a_p\right)dN_p. \tag{2.4.39}$$

One can eliminate also the gain by rearranging the steady-state equation (2.4.21), leading to

$$\frac{1}{\tau_p} - \Gamma v_g g = \frac{\Gamma R'_{\mathrm{sp}}}{N_p}. \tag{2.4.40}$$

Applying (2.4.40) one can rewrite (2.4.38) and (2.4.39) in a more compact matrix form

$$\frac{d}{dt}\begin{bmatrix} dN \\ dN_p \end{bmatrix} = \begin{bmatrix} -\gamma_{NN} & -\gamma_{NP} \\ \gamma_{PN} & -\gamma_{PP} \end{bmatrix}\begin{bmatrix} dN \\ dN_p \end{bmatrix} + \frac{\eta_i}{qV}\begin{bmatrix} dI \\ 0 \end{bmatrix}, \tag{2.4.41}$$

with

$$\gamma_{NN} = \frac{1}{\tau_{\Delta N}} + v_g a N_p, \tag{2.4.42}$$

$$\gamma_{NP} = v_g g - v_g a_p N_p = \frac{1}{\Gamma\tau_p} - \frac{R'_{\mathrm{sp}}}{N_p} - v_g a_p N_p, \tag{2.4.43}$$

$$\gamma_{PN} = \frac{\Gamma}{\tau'_{\Delta N}} + \Gamma v_g a N_p, \tag{2.4.44}$$

$$\gamma_{PP} = \frac{1}{\tau_p} - \Gamma v_g g + \Gamma v_g a_p N_p = \frac{\Gamma R'_{\mathrm{sp}}}{N_p} + \Gamma v_g a_p N_p. \tag{2.4.45}$$

Well above threshold, where photon density N_p is large enough, one can suppress several terms in (2.4.42–2.4.45) leading to

$$\gamma_{NN} = \frac{1}{\tau_{\Delta N}} + v_g a N_p, \tag{2.4.46}$$

$$\gamma_{NP} = \frac{1}{\Gamma\tau_p} - v_g a_p N_p, \tag{2.4.47}$$

$$\gamma_{PN} = \Gamma v_g a N_p, \tag{2.4.48}$$

$$\gamma_{PP} = \Gamma v_g a_p N_p. \tag{2.4.49}$$

By investigating solutions of the equation set (2.4.41) one can get insight into the physical processes in a semiconductor laser under small-signal modulation. The coefficients (2.4.42–2.4.45) or (2.4.46–2.4.49) can be considered as constants. Although these coefficients depend on the carrier and photon densities N and N_p, these densities are determined at a certain working current I and can be extracted by solving the equation set (2.4.12–2.4.17) for the steady-state case, as was described earlier.

Important are the small-signal responses to a sinusoidal current modulation dI. To solve (2.4.41) for this case, one can assume sinusoidal time dependence of all three small-signal variables dI, dN and dN_p:

$$dI(t) = I_1 e^{j\omega t}, \tag{2.4.50}$$

$$dN(t) = N_1 e^{j\omega t}, \tag{2.4.51}$$

$$dN_p(t) = N_{p1} e^{j\omega t}, \tag{2.4.52}$$

where I_1, N_1 and N_{p1} are constant complex amplitudes (including magnitude and phase) of the modulation current, carrier density and photon density, accordingly, and j is the complex unity. Introducing (2.4.50–2.4.52) into (2.4.41), and after some rearrangements one obtains the following equation set for the complex amplitudes N_1 and N_{p1}:

$$\begin{bmatrix} \gamma_{NN} + j\omega & \gamma_{NP} \\ -\gamma_{PN} & \gamma_{PP} + j\omega \end{bmatrix} \begin{bmatrix} N_1 \\ N_{p1} \end{bmatrix} = \frac{\eta_i I_1}{qV} \begin{bmatrix} 1 \\ 0 \end{bmatrix}. \tag{2.4.53}$$

To get the solutions one needs to obtain the determinant of the matrix from (2.4.53), which is given by

$$\Delta \equiv \begin{vmatrix} \gamma_{NN} + j\omega & \gamma_{NP} \\ -\gamma_{PN} & \gamma_{PP} + j\omega \end{vmatrix} = \gamma_{NP}\gamma_{PN} + \gamma_{NN}\gamma_{PP} - \omega^2 + j\omega(\gamma_{NN} + \gamma_{PP}). \tag{2.4.54}$$

The small-signal carrier and photon densities can be consequently written down as

$$N_1 = \frac{\eta_i I_1}{qV} \cdot \frac{\gamma_{PP} + j\omega}{\omega_R^2} \cdot H(\omega), \tag{2.4.55}$$

$$N_{p1} = \frac{\eta_i I_1}{qV} \cdot \frac{\gamma_{PN}}{\omega_R^2} \cdot H(\omega), \tag{2.4.56}$$

with the two-parameter modulation transfer function $H(\omega)$ given by

$$H(\omega) = \frac{\omega_R^2}{\Delta} \equiv \frac{\omega_R^2}{\omega_R^2 - \omega^2 + j\omega\gamma}. \tag{2.4.57}$$

Here the relaxation resonance frequency ω_R and the damping factor γ are defined as follows:

$$\omega_R^2 \equiv \gamma_{NP}\gamma_{PN} + \gamma_{NN}\gamma_{PP}, \tag{2.4.58}$$

$$\gamma \equiv \gamma_{NN} + \gamma_{PP}. \tag{2.4.59}$$

One can insert the coefficients (2.4.42–2.4.45) into (2.4.58) and (2.4.59) and make some rearrangements in order to get the relaxation resonance frequency and the damping factor in the terms of the laser parameters:

$$\omega_R^2 = \frac{N_p v_g a}{\tau_p} + \left(\frac{\Gamma N_p v_g a_p}{\tau_{\Delta N}} - \frac{\Gamma v_g g}{\tau_{\Delta N}}\right)\left(1 - \frac{\tau_{\Delta N}}{\tau'_{\Delta N}}\right) + \frac{1}{\tau_{\Delta N}\tau_p}, \tag{2.4.60}$$

$$\gamma = \left(\frac{1}{\tau_{\Delta N}} + N_p v_g a\right) + \left(\frac{1}{\tau_p} + \Gamma N_p v_g a_p\right) - \Gamma v_g g. \tag{2.4.61}$$

As one can see from the terms grouped together in (2.4.61), in the first bracket terms leading to losses in the carrier density caused by the changes in this carrier density itself are present. The second bracket represents terms, which lead to losses in the photon density caused by the changes in this photon density itself. Finally, the last term of this equation is the term corresponding to the increase of the photon density, caused by the changes in this photon density itself. In (2.4.60) many terms from (2.4.61) are present, but in a modified form, divided by the differential carrier lifetimes $\tau_{\Delta N}$ and $\tau'_{\Delta N}$ or photon lifetime τ_p. Eliminating gain with (2.4.40) one obtains

$$\omega_R^2 = \frac{v_g a N_p}{\tau_p} + \left(\frac{\Gamma v_g a_p N_p}{\tau_{\Delta N}} + \frac{\Gamma R'_{\mathrm{sp}}}{N_p \tau_{\Delta N}}\right)\left(1 - \frac{\tau_{\Delta N}}{\tau'_{\Delta N}}\right) + \frac{1}{\tau'_{\Delta N}\tau_p}, \tag{2.4.62}$$

$$\gamma = v_g a N_p\left(1 + \frac{\Gamma a_p}{a}\right) + \frac{1}{\tau_{\Delta N}} + \frac{\Gamma R'_{\mathrm{sp}}}{N_p}. \tag{2.4.63}$$

Equation 2.4.62 for the relaxation resonance frequency ω_R can be further simplified for the case that the laser operates well above threshold. In this case the first term will dominate and the relaxation resonance frequency can be expressed as

$$\omega_R^2 \approx \frac{v_g a N_p}{\tau_p}. \tag{2.4.64}$$

The damping factor γ (2.4.63) can be rewritten using (2.4.64) for the laser operation well above threshold to the following equation:

$$\gamma \approx K f_R^2 + \gamma_0, \tag{2.4.65}$$

where

$$f_R = \frac{\omega_R}{2\pi}, \tag{2.4.66}$$

and the K-factor K with the damping factor offset γ_0 are defined as follows:

$$K = 4\pi^2 \tau_p \left(1 + \frac{\Gamma a_p}{a}\right), \tag{2.4.67}$$

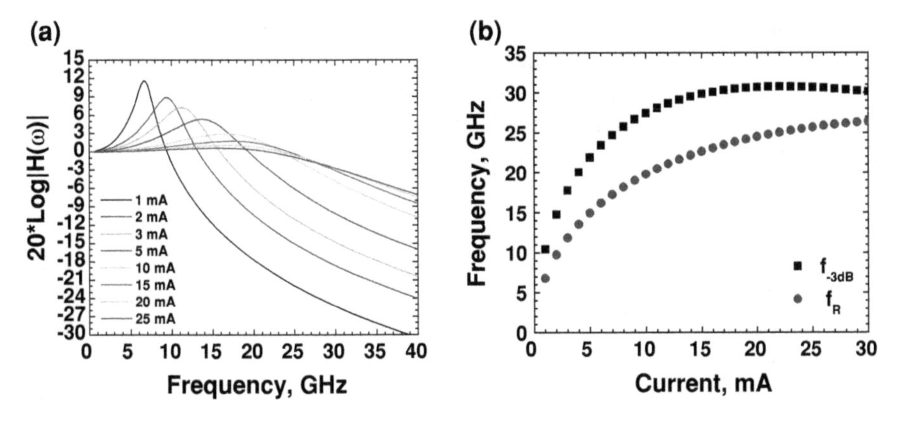

Fig. 2.39 Simulated modulation transfer functions for different currents (**a**) and corresponding relaxation resonance frequency and bandwidths (**b**) for an oxide-confined VCSEL

$$\gamma_0 = \frac{1}{\tau_{\Delta N}} + \frac{\Gamma R'_{sp}}{N_p}. \tag{2.4.68}$$

In Fig. 2.39, a simulated modulation transfer functions for several different currents for an oxide-confined VCSEL according to (2.4.57) are shown.

The modulation transfer function can be expressed in decibels (dB). Because not the optical output power of the laser but rather the electrical power on the photodetector, which corresponds to the squared optical power, is relevant for applications, conversion to decibels is carried out using a factor of 20 rather than 10 before logarithm, as shown in the last term in the equation below:

$$H[\mathrm{dB}] = 10 * \log|\mathrm{H}|^2 = 20 * \log|\mathrm{H}|. \tag{2.4.69}$$

One can see from Fig. 2.39a that at low currents relaxation resonance is well-marked, but at higher currents it becomes damped because of the increased damping factor according to (2.4.65). The position of the relaxation resonance also shifts to larger frequencies at larger currents, which can be more clearly seen from the Fig. 2.39b, where corresponding values for the relaxation resonance frequency and bandwidth are shown. One can understand the shift of the relaxation resonance frequency with current by combining (2.4.18) with (2.4.29) and inserting it into (2.4.64), leading together with (2.4.66) to

$$f_R = \frac{1}{2\pi}\sqrt{\frac{\eta_i v_g a}{qV_p}} \cdot \sqrt{I - I_{\mathrm{th}}} = D\sqrt{I - I_{\mathrm{th}}}, \tag{2.4.70}$$

where the D-factor D, characterizing the slope of the relaxation resonance frequency with current, was introduced. The D-factor is in a common case not constant because of the dependence of the differential gain a on carrier and photon densities according to (2.4.36). At higher carrier and photon densities it becomes smaller leading to a weaker increase of the relaxation resonance frequency. Also temperature effects, which were not considered in this section, play an important role. One defines empirically the corresponding modulation current efficiency factor M for the 3 dB-frequency $f_{\text{-3dB}}$, also called the bandwidth of the laser, which is the frequency, at which the modulation transfer function decreases by 3 dB compared to the starting point at zero frequency:

$$f_{-3\,\mathrm{dB}} = M\sqrt{I - I_{\mathrm{th}}}. \tag{2.4.71}$$

The linear increase of the relaxation resonance frequency f_R and of the 3 dB-frequency $f_{\text{-3dB}}$ with the square root of the current above threshold at lower currents is demonstrated in Fig. 2.40a, where corresponding values for an oxide-confined VCSEL are shown.

In Fig. 2.40b extracted dependence of the damping factor on the squared relaxation resonance frequency according to (2.4.65) is shown. Again, because the K-factor K (2.4.67) and the damping offset γ_0 (2.4.68) are commonly not constant but depend on carrier and photon densities, the linear relationship is not maintained over the whole range. At larger relaxation resonance frequencies, which correspond to larger operation currents and thus larger carrier and photon densities, both K and γ_0 increase, leading to a super linear dependence. Again thermal effects would play also here a very important role, making the trends more pronounced.

One can rewrite (2.4.65) using (2.4.70), which will lead to a linear relationship between the damping factor and the operation current:

$$\gamma \approx KD^2(I - I_{\mathrm{th}}) + \gamma_0. \tag{2.4.72}$$

From the measurement point of view one can get access to the modulation transfer function H by measuring the modulation of the laser output power P_1, leaving the small modulation of the driving current I_1 constant and making the

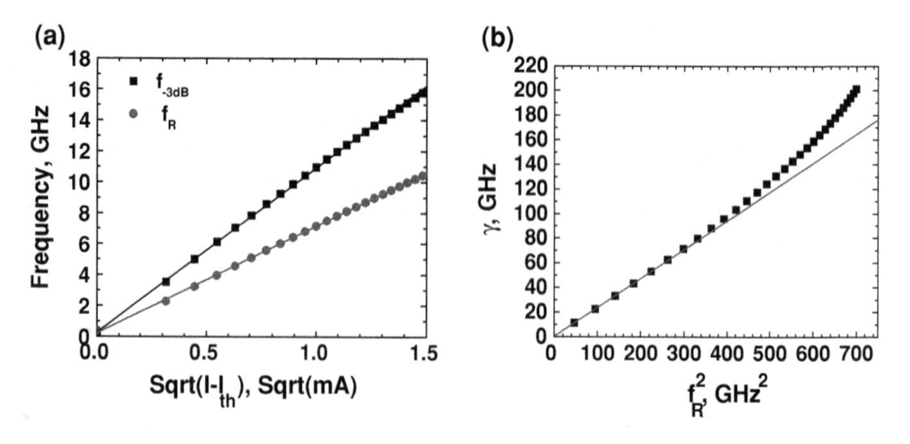

Fig. 2.40 Relaxation resonance frequency and 3 dB-frequency as functions of the square root of the current above threshold (**a**) and damping factor as a function of the squared relaxation resonance frequency (**b**) with corresponding linear fits for an oxide-confined VCSEL

frequency sweep. Using (2.4.56) together with (2.4.48) and (2.4.64) for the operation well above threshold one obtains

$$\frac{N_{p1}}{I_1} = \frac{\eta_i}{qV}\Gamma\tau_p H(\omega). \tag{2.4.73}$$

Now applying (2.4.18) and (2.4.11) one can expand (2.4.73) further to

$$\frac{P_1}{I_1} = \eta_i\eta_0\frac{hv}{q}H(\omega), \tag{2.4.74}$$

with the photon energy hv. As one can see from (2.4.74), the modulation transfer function can be measured experimentally, giving the possibility to get access to important laser parameters. Of course one should pay attention to other physical effects not considered in this section, e.g. thermal effects or electrical parasitics. For the latter one can introduce an additional term into (2.4.57), which, after some rearrangements, using (2.4.66) can be represented by

$$H(f) = \frac{f_R^2}{f_R^2 - f^2 + j\cdot f\cdot\frac{\gamma}{2\pi}}\cdot\frac{1}{1 + j\cdot\frac{f}{f_p}}. \tag{2.4.75}$$

The last term in (2.4.75) represents a low pass built by electrical parasitics of the laser, which were described in the previous sections. This low pass is characterized by its cut-off frequency f_p.

The modulation transfer function (2.4.75) has three parameters: the relaxation resonance frequency f_R, the damping factor γ and the cut-off frequency of electrical parasitics f_p. Accordingly, there are three types of limitations for the high speed operation of a semiconductor laser.

The first limitation is caused by the thermal effects, in fact by the internal laser heating, leading to an increase of the temperature of the active region. This is the so-called thermal limitation. Because temperature affects nearly each of the laser parameters used in the rate equations, the relaxation resonance frequency saturates at some current and starts to decrease at larger currents. Looking at (2.4.70) one can say, that the D-factor and the threshold current are temperature dependent, and the D-factor decreases with temperature, while the threshold current increases. Thus there is a limit for the relaxation resonance frequency, and it becomes limited by some maximum value $f_{R,\max}$. The maximum bandwidth that can be achieved in this case, assuming negligible damping and electrical parasitics, is

$$f_{-3\mathrm{dB,thermal}} \approx \sqrt{1+\sqrt{2}}\cdot f_{R,\max} \approx 1.55\cdot f_{R,\max}. \tag{2.4.76}$$

The second limitation is the damping limit, also called internal limitation. It arises from the fact, that the damping factor increases with the squared relaxation resonance frequency (2.4.65), while the bandwidth increases only approximately linear to it. Starting from some point, increase in the damping factor overcomes the increase in the relaxation resonance frequency, and the bandwidth of the laser begins to decrease. The maximum bandwidth possible in this case, assuming negligible electrical parasitics, is determined by the K-factor (2.4.67) and is

$$f_{-3\mathrm{dB,dampingl}} = \sqrt{2}\frac{2\pi}{K} \approx \frac{8.89}{K}. \tag{2.4.77}$$

The third type of limitations is caused by the presence of electrical parasitic elements inside of the laser, mostly parasitic resistances and capacitances. These electrical parasitics build a low pass, preventing high speed operation. With a given cut-off frequency of electrical parasitics f_p, maximum achievable bandwidth for an otherwise ideal case of perfect matching between relaxation resonance frequency and damping can be written down as

$$f_{-3\mathrm{dB,parasitics}} = \left(2+\sqrt{3}\right)f_p \approx 3.73 \cdot f_p. \tag{2.4.78}$$

The main goal of a researcher designing high speed VCSELs is to overcome these limitations. Commonly there are two or even all three types of limitations, which prohibit high speed laser operation. Consequently, one must apply many different concepts and very often some compromises should be met. Additionally, practical requirements like laser reliability, scalable and straightforward growth, fabrication and testing processes etc. should be satisfied.

References

1. Chuang SL (2009) Physics of photonic devices. Wiley, Hoboken
2. Coldren LA, Corzine SW (1995) Diode lasers and photonic integrated circuits. Wiley, New York
3. Kapon E (1999) Semiconductor lasers. Academic Press, San Diego
4. Cheng J, Dutta NK (2000) Vertical-cavity surface-emitting lasers: technology and applications. Gordon and Breach Science Publishers, Amsterdam
5. Crosslight, Crosslight Software Inc., http://www.crosslight.com/
6. PICS3D: photonic integrated circuit simulator in 3D, product description, http://www.crosslight.com/products/pics3d_mini_brochure.pdf
7. Silvaco, Silvaco Inc., http://www.silvaco.com/
8. WIAS TeSCA, Weierstraß-Institut für Angewandte Analysis und Stochastik, http://www.wias-berlin.de/software/tesca/index.html.de
9. Streiff M, Witzig A, Pfeiffer M, Royo P, Fichtner W (2003) A comprehensive VCSEL device simulator. IEEE J Sel Top Quantum Electron 9(3):879–891
10. Hadley GR (1995) Effective index model for vertical-cavity surface-emitting lasers. Opt Lett 20:1483–1485
11. Wenzel H, Wünsche HJ (1997) The effective frequency method in the analysis of vertical-cavity surface-emitting lasers. IEEE J Quantum Electron 33:1156–1162
12. Vukusic JA, Martinsson H, Gustavsson JS, Larsson A (2001) Numerical optimization of the single fundamental mode output from a surface modified vertical cavity surface emitting laser. IEEE J Quantum Electron 37:108–117
13. Klein B, Register LF, Hess K, Deppe DG, Deng Q (1998) Self-consistent Green's function approach to the analysis of dielectrically apertured vertical-cavity surface-emitting lasers. Appl Phys Lett 73(23):3324–3326

14. Noble MJ, Loehr JP, Lott JA (1998) Analysis of microcavity VCSEL lasing modes using a full-vector weighted index method. IEEE J Quantum Electron 34(10):1890–1903
15. Bienstman P, Baets R (2001) Optical modeling of photonic crystals and VCSEL's using eigenmode expansion and perfectly matched layers. Opt Quantum Electron 33(4/5):327–341
16. Bienstman P (2000–2001) Rigorous and efficient modelling of wavelength scale photonic components. Dissertation, Gent University, Faculty for Information Technologies, academic year 2000–2001
17. Bienstman P, Baets R, Vukusic J, Larsson A, Noble MJ, Brunner M, Gulden K, Debernardi P, Fratta L, Bava GP, Wenzel H, Klein B, Conradi O, Pregla R, Riyopoulos SA, Seurin J-FP, Chuang SL (2001) Comparison of optical VCSEL models on the simulation of oxide-confined devices. IEEE J Quantum Electron 37(12):1618–1631
18. CAMFR: Cavity Modelling Framework, http://camfr.sourceforge.net/
19. Itoh T (1989) Numerical techniques for microwave and millimeter-wave passive structures. Wiley, New York
20. Taflove A (1995) Computational electrodynamics, the finite-difference time-domain method. Artech House, Norwood
21. Morishita K (1983) Hybrid modes in circular cylindrical optical fibers. IEEE Trans Microw Theory Tech MTT-31(4):344–350
22. Ibanesen M, Johnson SG, Soljacic M, Joannopoulos JD, Fink Y, Weisberg O, Engness TD, Jacobs SA, Skorobogatiy M (2003) Analysis of mode structure in hollow dielectric waveguide fibers. Phys Rev E 67:046608
23. Aghaie KZ, Dangui V, Digonnet MJF, Fan S, Kino GS (2009) Classification of the core modes of hollow-core photonic-bandgap fibers. IEEE J Quantum Electron 45(9):1192–1200
24. Leisher PO, Chen C, Sulkin JD, Alias MSB, Sharif KAM, Choquette KD (2007) High modulation bandwidth implant-confined photonic crystal vertical-cavity surface-emitting lasers. IEEE Photonics Technol Lett 19(19):1541–1543
25. Danner AJ, Raftery JJ Jr, Leisher PO, Choquette KD (2006) Single mode photonic crystal vertical cavity lasers. Appl Phys Lett 88:091114
26. Song D-S, Kim S-H, Park H-G, Kim C-K, Lee Y-H (2002) Single-fundamental-mode photonic-crystal vertical-cavity surface-emitting lasers. Appl Phys Lett 80(21):3901–3903
27. Song D-S, Lee Y-J, Choi H-W, Lee Y-H (2003) Polarization-controlled, single-transverse-mode, photonic-crystal, vertical-cavity, surface-emitting lasers. Appl Phys Lett 82(19):3182–3184
28. Yang HPD, Chang YH, Lai FI, Yu HC, Hsu YJ, Lin G, Hsiao RS, Kuo HC, Wang SC, Chi JY (2005) Singlemode InAs quantum dot photonic crystal VCSELs. Electron Lett 41(20):1130–1132
29. Leisher PO, Danner AJ, Raftery JJ Jr, Siriani D, Choquette KD (2006) Loss and index guiding in single-mode proton-implanted holey vertical-cavity surface-emitting lasers. IEEE J Quantum Electron 42(10):1091–1096
30. Martinsson H, Vukusic JA, Grabherr M, Michalcik R, Jäger R, Ebeling KJ, Larsson A (1999) Transverse mode selection in large-area oxide-confined vertical-cavity surface-emitting lasers using a shallow surface relief. IEEE Photonics Technol Lett 11(12):1526–1538
31. Haglund A, Gustavsson JS, Vukusic J, Modh P, Larsson A (2004) Single fundamental-mode output power exceeding 6 mW from VCSELs with shallow surface relief. IEEE Photonics Technol Lett 16(2):368–370
32. Söderberg E, Gustavsson JS, Modh P, Larsson A, Zhang Z, Berggren J, Hammar M (2007) Suppression of higher order transverse and oxide modes in 1.3 μm InGaAs VCSELs by an inverted surface relief. IEEE Photonics Technol Lett 19(5):327–329
33. Mutig A (2004) Entwicklung von Oberflächenemittierenden Lasern. Diploma work, Technical University of Berlin, Institute of Solid State Physics, 3 May 2004
34. Stinavasan K, Borselli M, Painter O, Stintz A, Krishna S (2006) Cavity Q, mode volume, and lasing threshold in small diameter AlGaAs microdisks with embedded quantum dots. Opt Express 14(3):1094–1105
35. Piprek J (2003) Semiconductor optoelectronic devices. Elsevier Science, San Diego

36. Nextnano++, Walter Schottky Institut, Technische Universität München, Germany, http://www.wsi.tum.de/nextnanoplus/
37. Chang Y-C (2008) Engineering vertical-cavity surface-emitting lasers for high-speed operation. Dissertation, University of California, Santa Barbara, Dec 2008
38. Suzuki N, Anan T, Hatakeyama H, Fukatsu K, Tokutome K, Akagawa T, Tsuji M (2009) High speed 1.1 μm-range InGaAs-based VCSELs. IEICE Trans Electron E92-C(7):942–950
39. Al-Omari AN, Al-Kofahi IK, Lear K (2009) Fabrication, performance and parasitic parameter extraction of 850 nm high-speed vertical-cavity lasers. Semiconduct Sci Technol 24(9):095024 8 pp
40. Ou Y, Gustavsson JS, Westbergh P, Haglund A, Larsson A, Joel A (2009) Impedance characteristics and parasitic speed limitations of high speed 850 nm VCSELs. IEEE Photonics Technol Lett 21(24):1840–1842
41. Chang Y-C, Coldren LA (2009) Efficient, high-data-rate, tapered oxide-aperture vertical-cavity surface-emitting lasers. IEEE J Sel Top Quantum Electron 15(3):704–715
42. Hewlett-Packard Test and Measurement Application Note 95-1, S-parameter techniques for faster, more accurate network design, http://www.hp.com/go/tmappnotes, Hewlett-Packard Company, Palo Alto, CA, USA, 1997
43. Microwave Office, AWR, http://web.awrcorp.com/Usa/Products/Microwave-Office/
44. AppCAD, Avago Technologies, http://www.hp.woodshot.com/
45. TXLine, AWR, http://web.awrcorp.com/Usa/Products/Optional-Products/TX-Line/
46. Al-Omari AN, Carey GP, Hallstein S, Watson JP, Dang G, Lear KL (2006) Low thermal resistance high-speed top-emitting 980 nm VCSELs. IEEE Photonics Technol Lett 18(11):1225–1227
47. Satuby Y, Orenstein M (1999) Mode-coupling effects on small-signal modulation of multitransverse-mode vertical-cavity semiconductor lasers. IEEE J Quantum Electron 35(6): 944–954
48. Zei L-G, Ebers S, Kropp J-R, Petermann K (2001) Noise performance of multimode VCSELs. J Lightwave Technol 19(6):884–892
49. Westbergh P, Gustavsson JS, Haglund A, Sköld M, Joel A, Larsson A (2009) High-speed, low-current-density 850 nm VCSELs. IEEE JSTQE 15(3):694–703

Chapter 3
VCSEL Growth and Fabrication

Creation of high-quality VCSEL devices commonly requires three types of activities, which are laser designing process, wafer growth and device fabrication. These activities are not independent from each other but very closely interconnected, as shown in Fig. 3.1. For a proper laser design clear understanding of the growth process as well as explicit knowledge of the fabrication process are indispensable. Accuracy and tolerances of each growth and fabrication step should be considered already during the designing phase. And vice versa, knowledge of critical points in growth and fabrication can be applied to develop a stable design which will avoid possible problems. A VCSEL design can be optimized in such a way that the overall complexity of the whole manufacturing process is minimized. Some critical issues in growth can be avoided by applying additional fabrication steps and vice versa.

After the VCSEL designing process was considered in the previous chapter, VCSEL growth and fabrication will be described in the present chapter. To obtain correct measurement results at high frequencies and to interpret them properly are additional decisive not trivial tasks in the laser research process. Measurement results and analysis of 980 nm SML-VCSELs, 980 nm QW-VCSELs and 850 nm QW-VCSELs manufactured within the framework of the present thesis would be presented in the following chapters.

3.1 Growth of the VCSEL Epitaxial Structure

Epitaxial growth is a very critical and complex process. High quality semiconductor material resulting from epitaxial growth is an absolutely indispensable precondition for proper laser manufacturing. There are two mature technologies, which have established themselves for growth of high quality VCSEL wafers: the metal–organic chemical vapour deposition (MOCVD), sometimes called also metal–organic vapour phase epitaxy (MOVPE), and the molecular beam epitaxy

A. Mutig, *High Speed VCSELs for Optical Interconnects*, Springer Theses,
DOI: 10.1007/978-3-642-16570-2_3,

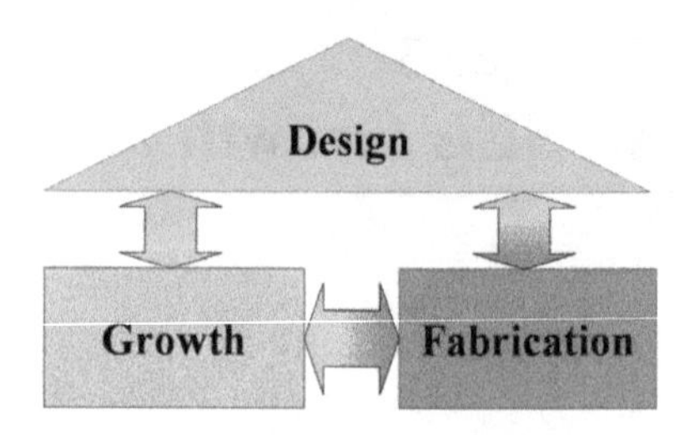

Fig. 3.1 Interaction of the laser designing, wafer growth and device fabrication processes

(MBE). The growth process in an MOCVD reactor results from the reaction of the products, created when the growth precursors decompose on contact with the hot substrate. The main growth control parameters are gas fluxes, substrate temperature, reactor pressure and growth time. In an MBE reactor growth materials are contained separately in effusion cells in elemental form, and molecules evaporate off from the cells under high vacuum condition bringing epitaxial material to the substrate. The flux of molecules is controlled by varying the cell temperatures and operating the cell shutters.

Each of these epitaxial growth technologies has their advantages [1]. One of the main advantages of the MOCVD is the higher growth rate, making this technology more suitable for large scale VCSEL mass production. Because gas fluxes can be varied continuously, also compositional gradings can be easily realized. One of the main advantages of the MBE is the more precisely control of the layer thicknesses, especially for very thin layers. Because of the limited number of the effusion cells in an MBE reactor, only a relatively small number of compositions is possible within one growth process. Compositional gradings are realized in MBE by deposition of two compositions with different thicknesses, as demonstrated in Fig. 3.2 for an example of a 10 nm thick AlGaAs linear grading from pure GaAs to $Al_{0.90}Ga_{0.10}As$. Such gradings are called digital gradings.

In the present work VCSELs grown with both MBE and MOCVD were fabricated and investigated. Because of the better thickness control, which is very important for the growth of the SML active region [2], MBE technology was applied for the epitaxial growth of 980 nm SML-VCSELs. The growth sequence for the SML InGaAs active region is schematically shown in Fig. 3.3, where small rectangles represent In-rich material.

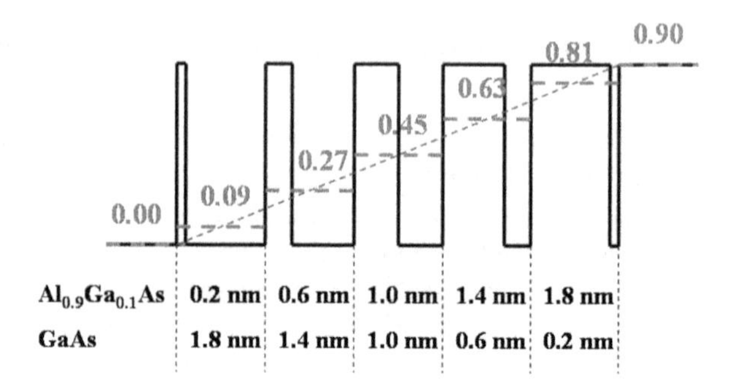

Fig. 3.2 AlGaAs 10 nm thick digital grading grown by MBE, applied in the DBR mirrors of the 980 nm SML-VCSELs investigated in this work

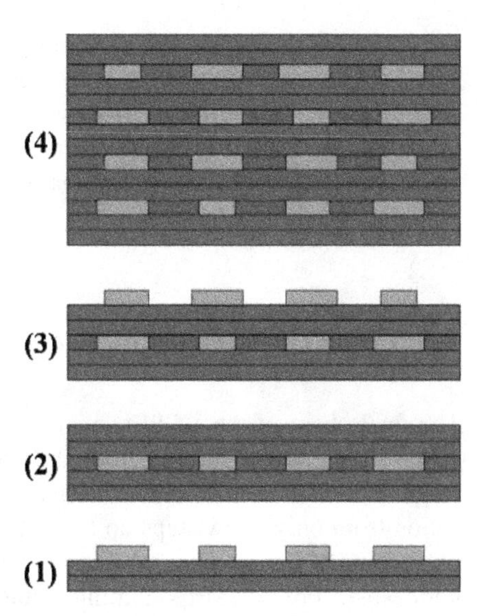

Fig. 3.3 Schematic illustration of the SML growth for the InGaAs active region; small rectangles represent In-rich material

First, on the GaAs matrix less than one monolayer of InAs is deposited, resulting in formation of one monolayer thick In-rich islands on the GaAs surface. These islands are overgrown in the next step with several monolayers of GaAs, leading to a flat surface. After that the second InAs deposition takes place. The growth of the islands in the third step is affected by the underlying structure, resulting in a vertical correlation between different In-rich layers. The whole procedure is repeated several times, leading to a high quality InGaAs active region.

For the 980 SML-VCSELs described in this work following parameters for the active region grown in the SML growth regime were applied: nominal growth temperature was 490°C, nominal thickness of the InAs submonolayers in a stack was 1 Å, nominal thickness of the GaAs layers between the InAs submonolayers in a stack was 6 Å, the number of the InAs submonolayers in one stack was 10. Active region of SML-VCSELs contained three SML-stacks, each consisting of 10 InAs submonolayers. Between SML-stacks 13 nm thick GaAs spacers were grown. Nominal emission wavelength was calibrated to be at 970 nm. Average In composition in a SML-stack was around 20%. For DBR mirrors digital gradings shown in Fig. 3.2 were applied. All compositional gradings in the MBE grown VCSELs were digital gradings.

For devices having QWs as active region MOCVD growth was applied, bringing VCSELs one step closer to the possible future commercialization. These were 980 nm QW-VCSELs and 850 nm QW-VCSELs. In these devices 20 nm thick gradings were utilized. In all devices fabricated and investigated in this work, both with MBE and MOCVD, carbon (C) was used as p-type dopant and silicon (Si) as n-type dopant.

Growth quality and emission wavelength of the active region was controlled by measuring of the photo luminescence (PL) on calibration structures. Overall growth quality of the whole VCSEL structure was inspected by investigation of the wafer surface in respect to possible growth defects. Defect density and cavity dip wavelength were strictly controlled. Fig. 3.4 shows as example a map of the cavity dip wavelength of the 980 nm SML-VCSEL wafer, measured after the MBE growth process. As one can see, more than 90% of the surface is within growth specifications, ensuring a proper growth process.

3.2 VCSEL Fabrication Technology

The final phase of the VCSEL manufacturing is the device fabrication process. It starts with a wafer coming from the epitaxy and commonly consists of several steps. The number of steps can vary depending on the complexity of the fabrication from only a few steps up to 15–20 steps or even more. Also the complexity of any individual step as well as of the whole fabrication procedure can vary noticeable. For VCSELs aiming future markets possibility of a large scale industrial production should be one of the important considerations. For this purpose it is of a great importance to hold the number of steps as well as their complexity limited. The overall fabrication process should be optimized in respect to integrity and simplicity to achieve high yield, proper laser performance, high device reliability, and to keep possible costs low. These considerations are also important for a researcher, since high-quality reliable devices coming from a high yield fabrication contribute deciding to a laser research and development process.

There are several competing factors, which affect laser performance and thus should be considered during the device designing process. These are thermal, internal and electrical properties already mentioned in the previous chapter.

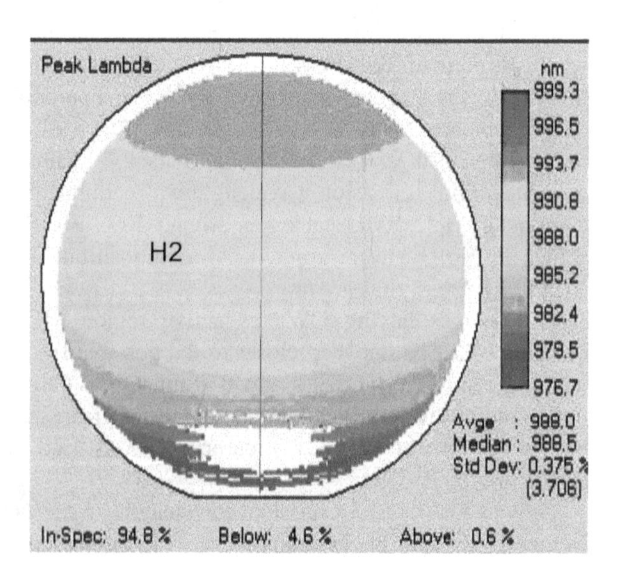

Fig. 3.4 Cavity dip wavelength of the 980 nm SML-VCSEL wafer

While small mesa diameters and deep mesa etching are needed to reduce electrical parasitics, from the thermal point of view rather larger mesas and shallow etching are preferable. Higher doping levels in the mirrors and in the cavity region reduce electrical resistance of the devices but also increase absorption losses. Small oxide aperture diameters reduce mode volume and threshold current, and this improves the high speed performance of the lasers, but this also increases current density and decreases output power. Thus the overall laser design incorporates a rather complex compromise of many competing factors, including high modulation speed, high thermal conductivity, large output power, and low current density for a better device reliability.

In the present work all VCSELs have been fabricated using the same optimized mask set and the same fabrication steps. However, because of different process parameters of several critical steps, first of all of the selective wet oxidation, VCSELs with different characteristics could be realized, ensuring laser performance progress not only by improvements of the epitaxial structure but also because of the optimizations in the fabrication process.

For the on-chip characterization the number of steps could be reduced, since no additional mechanical stability for bonding etc. is required. The optimized fabrication scheme for all VCSELs investigated in this work consists of eight steps and is shown in Fig. 3.5. The fabrication starts with an epitaxial wafer (Fig. 3.5a) and some cleaning steps, which were introduced also in the proper places of the following fabrication process. All fabrication steps were optimized for manufacturing of a quarter of a 3″-wafer, resulting in a relatively straightforward, very stable, homogeneous, highly reproducible and reliable process flow. The overall fabrication process yield on a quarter 3″-wafer was higher than 95%, providing many hundreds of VCSELs for a comprehensive detailed characterization.

After the initial cleaning top metal contact was deposited using electron beam evaporation system (Fig. 3.5b). All VCSELs described in the present work have p-type doped top DBR mirror. The bottom DBR mirror was n-type doped for all devices. The top p-type contact was a Ti/Pt/Au contact with the corresponding nominal metal thicknesses of 20/50/300 nm. In this step also the backside n-type contact have been evaporated, which was a Ni/AuGe/Au contact with 10/100/300 nm metal thicknesses. The AuGe layer was an alloy consisting of 88% Au and 12% Ge. The backside contact can be used for structures grown on n-type doped substrates for laser testing in the middle of the fabrication after the fifth step, and also for better control of the selective wet oxidation by utilizing calibration test structures, which can be electrically contacted and measured. In the case of undoped substrates the backside contact can not be used and can be skipped.

After the first metal deposition step the first mesa etching was performed in an inductive coupled plasma reactive ion etching (ICP-RIE) machine and Cl-based plasma process (Fig. 3.5c). The etching depth was controlled by the in situ reflectivity measurement system. As one example, in situ reflectivity measurements during the first and the second mesa etch of the 980 nm QW-VCSELs are shown in Fig. 3.6. Mirror pairs of the top and bottom DBR mirror as well as the cavity region can be clearly seen, enabling etch stop with a proper precision.

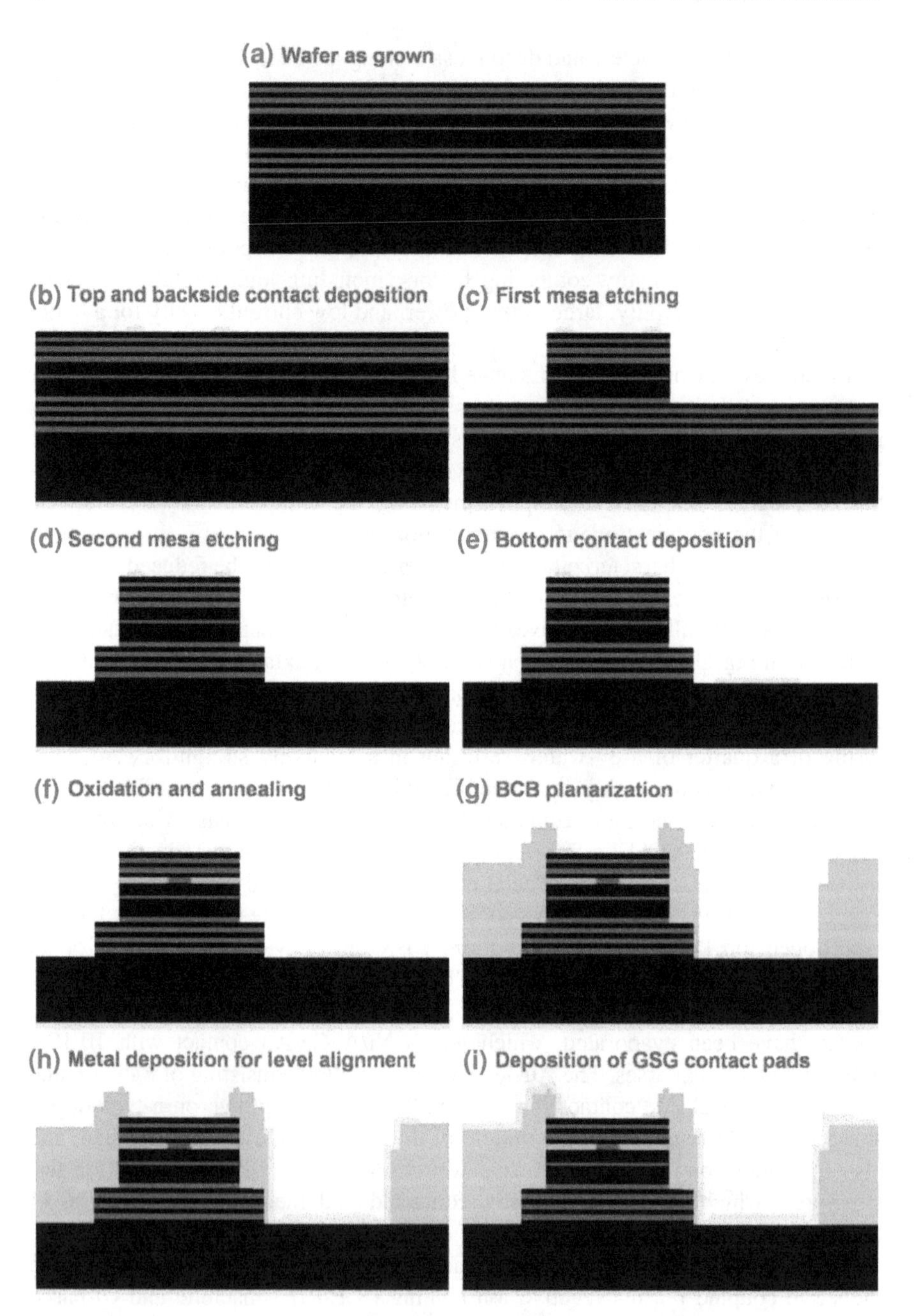

Fig. 3.5 VCSEL fabrication schema with individual steps. Position of the active region is shown in red, metal contacts are yellow, oxide aperture is light-blue and BCB is light-green. AlGaAs is shown in black and grey

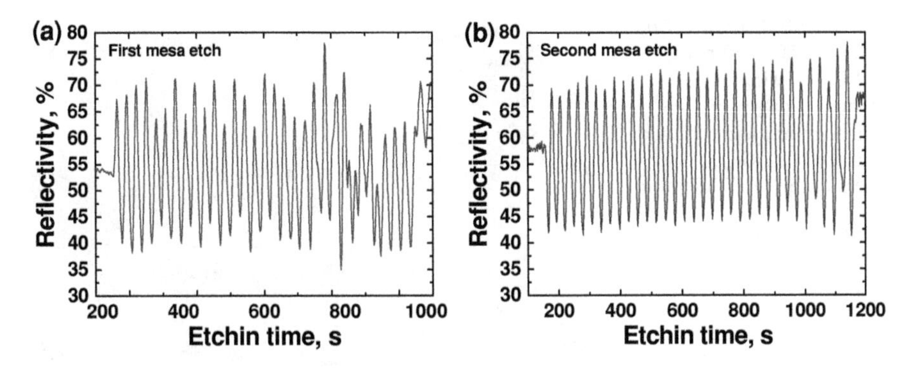

Fig. 3.6 In situ reflectivity measurements during the first (**a**) and the second (**b**) mesa etch of the 980 nm QW-VCSELs in the ICP-RIE reactor

The etching process was stopped after the cavity and a few of the bottom DBR mirror pairs have been exposed to ensure proper selective wet oxidation of the aperture layers with high Al content in the fifth step. The diameters of the first mesa varied nominally from 25 up to 36 μm in steps of 1 μm, giving the possibility to characterize up to twelve VCSELs with different oxide aperture diameters. The first mesa etching depth was in the range of ~4–6 μm, depending on the VCSEL structure.

The second mesa etching (Fig. 3.5d) was performed in the third step using the same machine and etching process. The presence of the second mesa aims to increase VCSEL thermal conductivity compared to the variant with only one deep mesa etching step. The radius of the second mesa is nominally 40 μm larger than the radius of the first mesa, improving heat flow from the active region to the substrate. At the same time the bottom n-type doped contact layer is exposed in this etching step.

As the next step n-type Ni/AuGe/Au bottom contact is deposited (Fig. 3.5e) using the same machine and process as in the first step for the backside contact. Both p-type top and n-type bottom contacts, evaporated in the first and fourth steps, can be contacted from the wafer surface, enabling further fabrication into design suitable for high speed on-wafer measurements.

After all the contacts have been deposited and the aperture layer with high Al content have been exposed during the first mesa etch, one can proceed with the formation of the oxide apertures by the selective wet oxidation of AlGaAs to Al_xO_y (Fig. 3.5f). The oxidation is carried out at temperatures between 400 and 420°C. Simultaneously the contacts are annealed. Because the oxidation rate is very sensitive to temperature and composition variations, special care is necessary during this step. Commonly variation of the oxidation rate leads to a difference of the oxide aperture diameters across the wafer piece for devices with the same first mesa diameter. This variation was in the range of 2–5 μm for a quarter piece of the 3″-wafer. Fig. 3.7 shows an SEM picture of the side view of a 850 nm QW-VCSEL, where two oxide apertures can be clearly seen as partially darken layers.

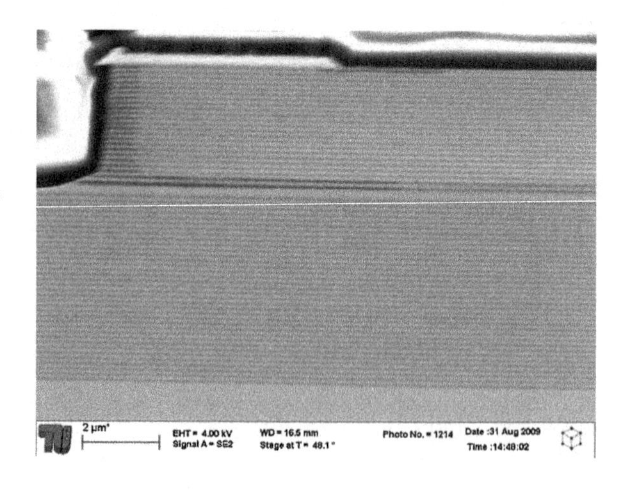

Fig. 3.7 SEM picture of the side view of a 850 nm QW-VCSEL showing oxide apertures (long partially darken layers) and partially oxidized $Al_{0.90}Ga_{0.10}As$ layers of the top DBR mirror

The sixth step was the optical lithography process utilizing thick bisbenzocyclobutene (BCB) layer for device planarization and reduction of parasitic contact pad capacitance (Fig. 3.5g). This is a very important step to overcome the limitation associated with electrical parasitics. BCB thickness depends thereby not only on the process parameters but also on the VCSEL geometry and geometry of surrounding structures. Thicknesses in the range of $\sim$8–11 µm have been used for devices investigated in this work, enabling reduction of the parasitic pad capacitance to values as low as only a few femtofarads.

Before the final metal deposition for the ground-signal-ground (GSG) high frequency contact pads an additional metal deposition step should be carried out to align the different height levels of the top p-type and bottom n-type contacts (Fig. 3.5h). For a better adhesion of the metal to BCB Cr/Au contacts were used with 50/300 nm thicknesses, respectively.

Finally, GSG high frequency Cr/Au contact pads were evaporated (Fig. 3.5i) using the same machine and process as in the seventh step. After this step fabricated VCSELs are ready for immediate CW and high frequency characterization.

Fig. 3.8 shows two SEM images of a fabricated VCSEL with the diameter of the first mesa equal to 32 µm. Hereby the entire VCSEL and a closer view are represented. Top and bottom contacts, GSG pad, two mesas and BCB layer can be clearly recognized.

Optical microscope images of a fabricated VCSEL array with different diameters of the first mesa and thus of the oxide aperture are shown in Fig. 3.9. The wafer can be cut between individual VCSELs, so that each VCSEL can be picked up from the array and placed into a package etc.

Summarized, a very stable, reliable, straightforward fabrication process with high yield and good uniformity was developed, aiming among other high VCSEL thermal conductivity, low electrical parasitics, high speed operation, on-wafer characterization and device reliability. The number and complexity of fabrication steps were minimized to optimize the process for a possible mass production and to reduce fabrication costs. Yields of 95% and higher were achieved, enabling

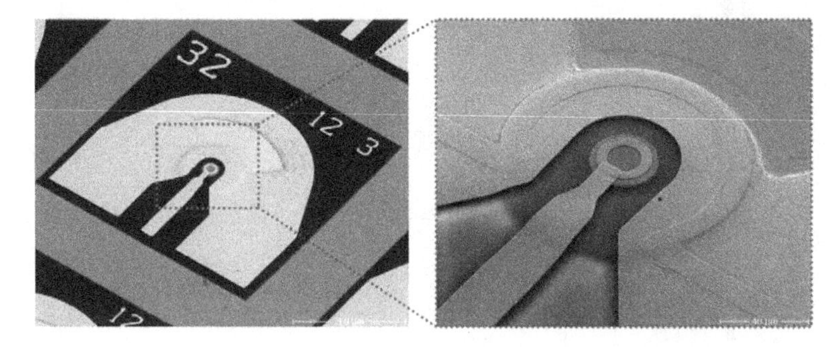

Fig. 3.8 SEM images of a fabricated VCSEL with the mesa diameter of 32 μm, showing the entire VCSEL and a closer view with contacts, two mesas, GSG pad and BCB layer

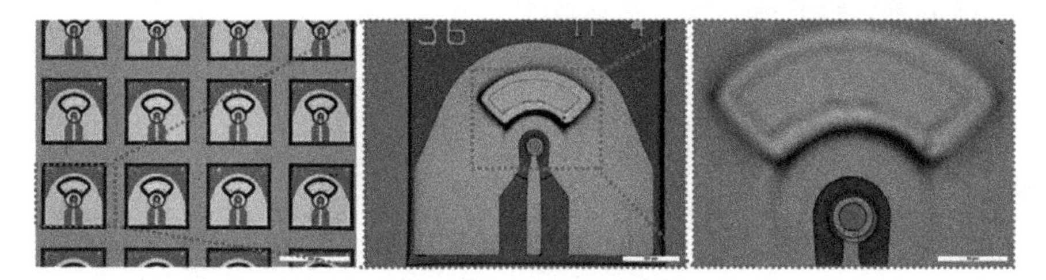

Fig. 3.9 Optical microscope images of a fabricated VCSEL array with different mesa and thus aperture diameters from 25 up to 36 μm

detailed characterization of fabricated VCSELs, which will be presented in the following chapters.

References

1. Sale TE (1995) Vertical cavity surface emitting lasers. Research Studies Press Ltd., Taunton
2. Krestnikov I, Ledentsov NN, Hoffmann A, Bimberg D (2001) Arrays of two-dimensional islands formed by submonolayer insertions: growth, properties, devices (review). Phys Stat Sol A 183(2):207–233

Chapter 4
High Temperature Stable 980 nm VCSEL Results

4.1 VCSELs with the SML Active Region

To achieve high temperature stability the whole VCSEL structure including active region, cavity region and DBR mirrors should be optimized. In the first generation of 980 nm VCSELs presented in this work one of the main optimization was the use of the active region grown in the submonolayer growth mode [4–8]. Depending on the growth conditions, active layers with carrier confinement centers caused by the composition and thickness variations can be fabricated [9]. These confinement centers increase high temperature stability of the active region by decreasing probability of thermal carrier escape processes. Also binary GaAs was applied in the optimized DBR mirrors, increasing their thermal conductivity. In the cavity region, AlGaAs layers with Al composition of 80% were utilized, increasing the thermal conductivity further. The overall design optimization in respect to high speed at high operating temperatures resulted in VCSELs operating at 20 Gbit/s at temperatures from 0 up to 120°C. These were to the best of our knowledge worldwide first VCSELs showing an open eye at 20 Gbit/s at 120°C.

4.1.1 Device Structure

The 980 nm SML-VCSELs were grown on the n-doped GaAs (100) substrate by MBE in order to achieve high quality, well controlled SML active region. The active region consisted of a triple stack of InGaAs short period superlattices (SPS), deposited in the submonolayer growth mode at 490°C to avoid reevaporation of In, with 13 nm thick GaAs spacers in between. Each InGaAs SPS was built up by SML deposition of 1 Å InAs covered by 6 Å GaAs, and repeating this procedure ten times. Thus at the end of the growth three stacks with 10 SML layers in each were present in the VCSEL structure, acting as active region. The SML growth in the present structure resulted in an inhomogeneous distribution of the In composition and a variation of the thickness of each InGaAs SPS between 6 and 8 nm,

A. Mutig, *High Speed VCSELs for Optical Interconnects*, Springer Theses,
DOI: 10.1007/978-3-642-16570-2_4, © Springer-Verlag Berlin Heidelberg 2011

found out by high resolution transmission electron microscopy and cross-sectional scanning tunneling microscopy investigations.

Cavity region consisted to a largest part of $Al_{0.80}Ga_{0.20}As$ and included Al rich aperture layers in the field intensity node for a later selective wet oxidation. The Al content of 80% used in the cavity layers provides a better thermal conductivity and reduces the mesa capacity due to the smaller dielectric constant compared to e.g. 30% Al. The aperture layers were a 12 nm thick binary AlAs surrounded by two 9.2 nm thick $Al_{0.90}Ga_{0.10}As$ layers. During the selective wet oxidation all these three layers, including a part of the surrounding $Al_{0.80}Ga_{0.20}As$ material, oxidized with different oxidation rates, leading to formation of a tapered Al_xO_y aperture, reducing optical scattering losses. The cavity length was 3/2 λ. In order to improve the temperature stability of the lasers, the cavity was red shifted from the peak gain at 25°C by nominally 15 nm.

To simplify device fabrication and avoid critical etching steps optimized doped $Al_{0.90}Ga_{0.10}As$/GaAs DBR mirrors were applied for both top and bottom mirrors. The top p-doped DBR contained 20.5 mirror pairs, while the bottom n-doped DBR consisted of 32.5 mirror pairs. Linear gradings with 10 nm thickness were utilized to improve electrical conductivity.

High speed devices with co-planar top contacts and different aperture diameters were fabricated using the fabrication process described in the previous chapter. All measured VCSELs have shown no noticeable degradation during the measurements. VCSELs with aperture diameters smaller than $\sim$3 μm have shown single-mode behavior while lasers with larger aperture diameters were multimode. To get a better understanding of physical processes and limiting factors of the fabricated lasers, three types of VCSELs have been investigated: single-mode VCSELs with aperture diameter $\sim$2 μm, multimode VCSELs with aperture diameter $\sim$5 μm and VCSELs, which had aperture diameter of $\sim$3 μm and were single-mode at lower currents but became multimode at higher current injection levels. In the following sections results for three typical devices with nominal aperture diameters of 2, 3 and 5 μm would be presented.

4.1.2 Static Characteristics

Proper static device characteristics are indispensable for a good high speed VCSEL operation. VCSELs with different aperture diameters naturally have different static and dynamic characteristics. As we will see in this section, there is an optimum value of the oxide aperture diameter, depending on the VCSEL structure, to achieve a good compromise between high output power, low threshold current, high differential efficiency, low electrical resistance, high thermal conductivity and other important static and dynamic laser characteristics, which in the most cases should be found experimentally.

In Fig. 4.1 L–I-curves and extracted values of the maximum differential efficiency and threshold current at different temperatures between 25 and 120°C for

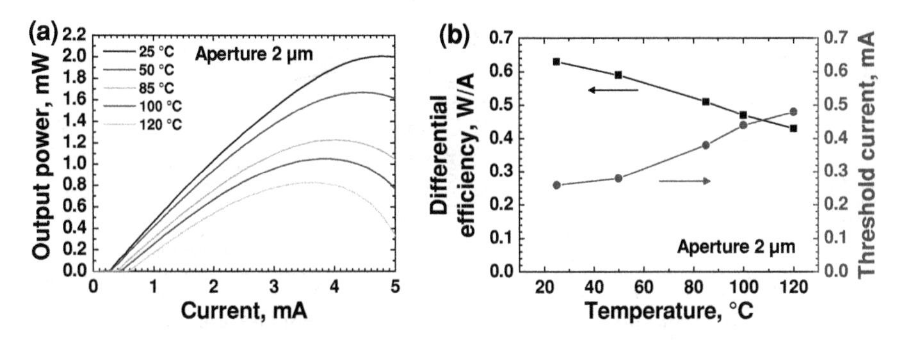

Fig. 4.1 L–I characteristics (a) and extracted values of the maximum differential efficiency and threshold current (b) at different temperatures between 25 and 120°C for the VCSEL with 2 μm aperture

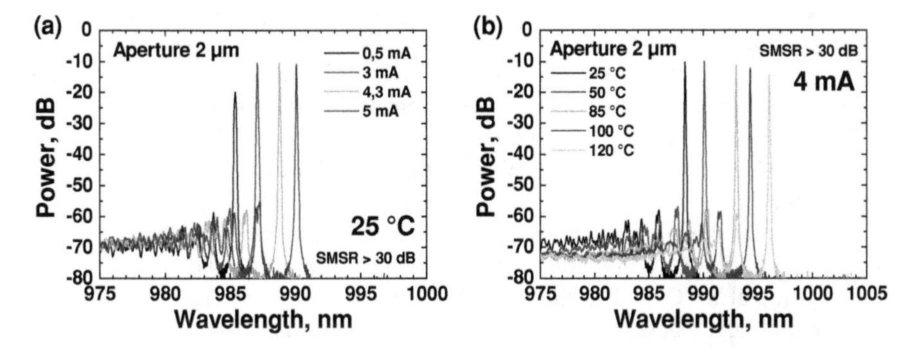

Fig. 4.2 Spectra at 25°C at different currents (a) and at 4 mA at different temperatures (b) for the VCSEL with 2 μm aperture

the single-mode VCSEL with 2 μm oxide aperture diameter are presented. The maximum output power of this device is ~2 mW at 25°C and reduces to ~1.2 mW at 85°C, and is still ~0.8 mW at 120°C. The maximum differential efficiency is hardly temperature dependent and reduces only from ~0.63 W/A at 25°C to ~0.51 W/A at 85°C, and is still ~0.43 W/A at 120°C. As one can see, both the output power and the peak differential efficiency are very temperature stable. The threshold current of the device is ~260 μA at 25°C and increases to ~380 μA at 85°C and further to ~480 μA at 120°C. The overall change of the threshold current from 25°C up to 120°C is smaller than by a factor of two, again showing its good temperature stability. The differential resistance of the VCSEL is ~380 Ω at current of 3 mA at temperature of 25°C, which is a quite high value and is caused by the small oxide aperture diameter of 2 μm.

The VCSEL stays single-mode in the whole current and temperature range with the side-mode suppression ratio (SMSR) larger than 30 dB, as shown in Fig. 4.2. Emission wavelength of the device is in the range of ~985 nm.

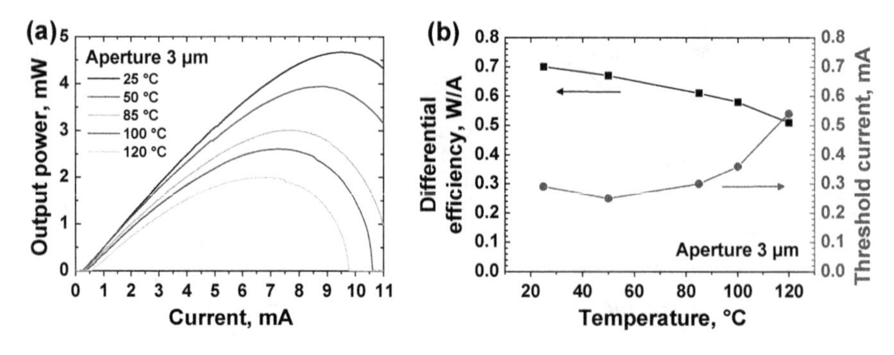

Fig. 4.3 L–I characteristics (**a**) and extracted values of the maximum differential efficiency and threshold current (**b**) at different temperatures between 25 and 120°C for the VCSEL with 3 µm aperture

Static device characteristics at different temperatures for the VCSEL with 3 µm aperture diameter are shown in Fig. 4.3. The maximum output power is temperature stable and reduces only from ~4.7 mW at 25°C to ~3 mW at 85°C, and further to ~2 mW at 120°C. The peak differential efficiency of the 3 µm VCSEL at 25°C is larger than the peak differential efficiency of the 2 µm VCSEL and is ~0.7 W/A compared to ~0.63 W/A for the 2 µm device. It decreases at higher temperatures to ~0.61 W/A at 85°C, and is still ~0.51 W/A at 120°C, but remains at all temperatures larger than that of the 2 µm VCSEL. The threshold current is ~290 µA at 25°C, decreases to ~250 µA at 50°C, and then increases to ~300 µA at 85°C, and further to ~540 µA at 120°C, showing a clear minimum around 50°C caused by the gain peak-cavity resonance detuning. This effect is not clearly observed for the single-mode VCSEL with 2 µm aperture diameter, because the increase of the gain due to the reduced detuning is overcompensated by the larger increase of the internal losses, which are temperature and also photon density dependent [10]. In addition small-aperture devices usually have a higher carrier density in the active region and confinement layers, making their internal efficiency more temperature sensitive. The differential resistance of the 3 µm VCSEL is ~200 Ω at 6 mA at 25°C.

In Fig. 4.4 L–I characteristics and values of maximum differential efficiency and threshold current at different temperatures for the VCSEL with 5 µm aperture diameter are shown. Logically this device has larger output power as compared to the previous two, which is ~8.4 mW at 25°C. It reduces to ~4.9 mW at 85°C, and is still ~3.2 mW at 120°C. The maximum differential efficiency is ~0.68 W/A at 25°C and reduces only to ~0.55 W/A at 85°C and further to ~0.45 W/A at 120°C. The gain–cavity detuning is clearly observed by taking a look at the threshold current as a function of temperature. The threshold current has the value of ~860 µA at 25°C, decreases to ~700 µA at 50°C, and then increases to ~790 µA at 85°C and further to ~1.22 mA at 120°C. The differential resistance of the 5 µm VCSEL is consequently lower as compared to the smaller aperture devices and has the value of ~92 Ω at 12 mA at 25°C.

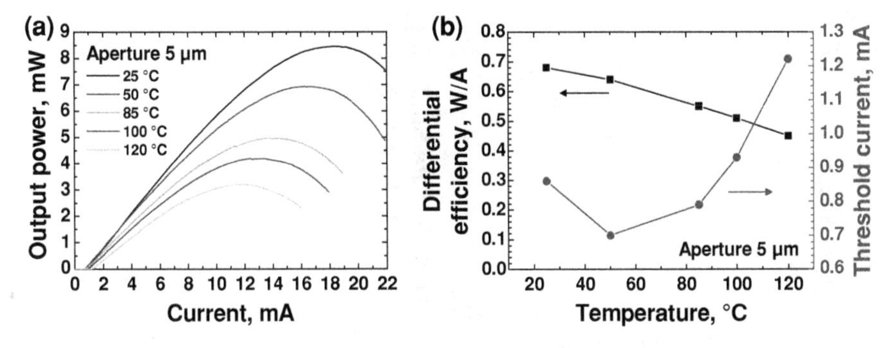

Fig. 4.4 L–I characteristics (**a**) and extracted values of the maximum differential efficiency and threshold current (**b**) at different temperatures between 25 and 120°C for the VCSEL with 5 μm aperture

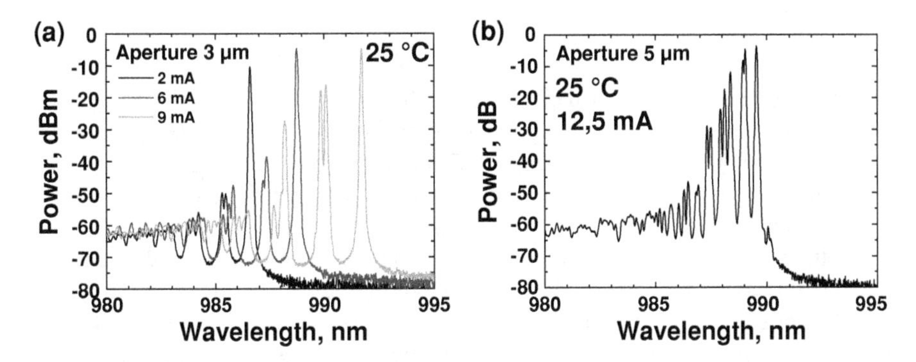

Fig. 4.5 Spectra of the 3 μm VCSEL at different currents (**a**) and of the 5 μm VCSEL at 12.5 mA (**b**) at 25°C

As one can see from Figs. 4.1, 4.3 and 4.4 the 3 μm VCSEL has the largest maximum differential efficiency at all temperatures as compared to the other two devices with 2 and 5 μm aperture diameter. Also the reduction of the maximum output power from 25 to 120°C has for this device the lowest value compared to another two VCSELs, which is a factor of ~2.4 for the 3 μm VCSEL, compared to the factor of ~2.5 for the 2 μm VCSEL and ~2.6 for the 5 μm VCSEL. Thus the 3 μm VCSEL appears to be the optimal laser for high speed operation, as will be confirmed during the high speed characterization.

Figure 4.5a shows emission spectra at 25°C at different current injection levels for the 3 μm VCSEL. It can observed, that in this laser a single-mode emission at lower currents converts to a multimode emission at higher currents. The 5 μm VCSEL operates in the whole current and temperature range multimode. Spectrum of the 5 μm VCSEL at 12.5 mA at 25°C is shown in Fig. 4.5b.

By measuring VCSEL emission wavelength as a function of the chuck temperature near the threshold, where self heating could be assumed to be negligible and the inner temperature of the device to be approximately equal to the chuck

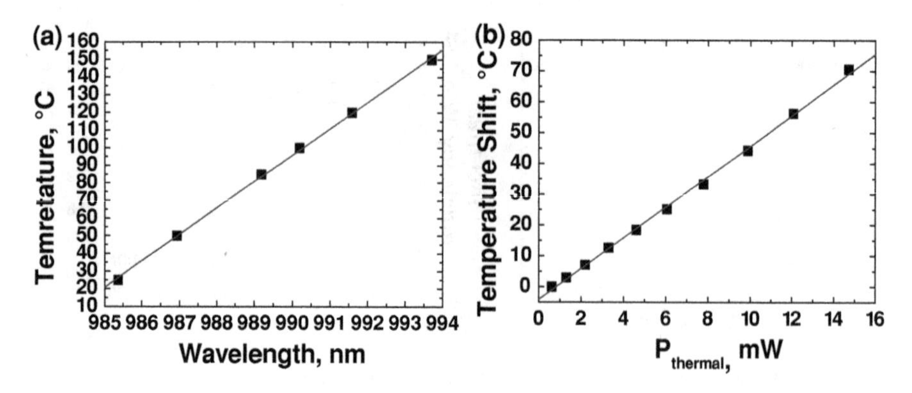

Fig. 4.6 Dependence of the VCSEL emission wavelength on the chuck temperature (**a**) and temperature shift as a function of the generated thermal power in a VCSEL (**b**) with corresponding linear fits for the VCSEL with aperture diameter of 2 μm

temperature, one can extract dependence of the VCSEL wavelength on the laser inner temperature. Thus, one obtains the possibility to extract the temperature of the active region from the measured VCSEL emission wavelength. Figure 4.6a shows the temperature as a function of the laser wavelength for the 2 μm VCSEL. A linear dependence can be easily recognized.

Plotting the difference between the laser inner temperature and the ambient temperature as a function of the thermal power, generated in the investigated VCSEL, one gets access to the thermal resistance of the device, which is defined by (2.3.8). The generated thermal power could be easily calculated from the measured L–U–I characteristics according to (2.3.9). In Fig. 4.6b the described dependence of the temperature shift on the generated thermal power is plotted again for the 2 μm VCSEL with the corresponding linear fit. Both linear fits in Fig. 4.6 represent the measured points very well. From the slope of the second fit the thermal resistance of the investigated device can be extracted directly, which has the dimensions of K/mW.

Following the described schema, thermal resistances of the three VCSELs were measured to be ~4.97, ~3.20 and ~2.10 K/mW for the 2, 3 and 5 μm devices, correspondingly. With increasing aperture diameter, thermal resistances decrease, as expected. Estimated inner device temperatures for different VCSELs as function of the driving current at a fixed chuck temperature of 25°C are shown in Fig. 4.7. Thermal rollover takes place at temperatures around ~100°C in the case of the smaller 2 μm VCSEL and at temperatures around ~120°C for the 3 and 5 μm VCSELs.

As can be seen from the measured static characteristics, all investigated VCSELs demonstrate good temperature stability in respect to the output power, differential efficiency and threshold current. Nevertheless the laser with the aperture diameter of 3 μm has a weaker decrease of the output power with temperature and a larger differential efficiency at all temperatures.

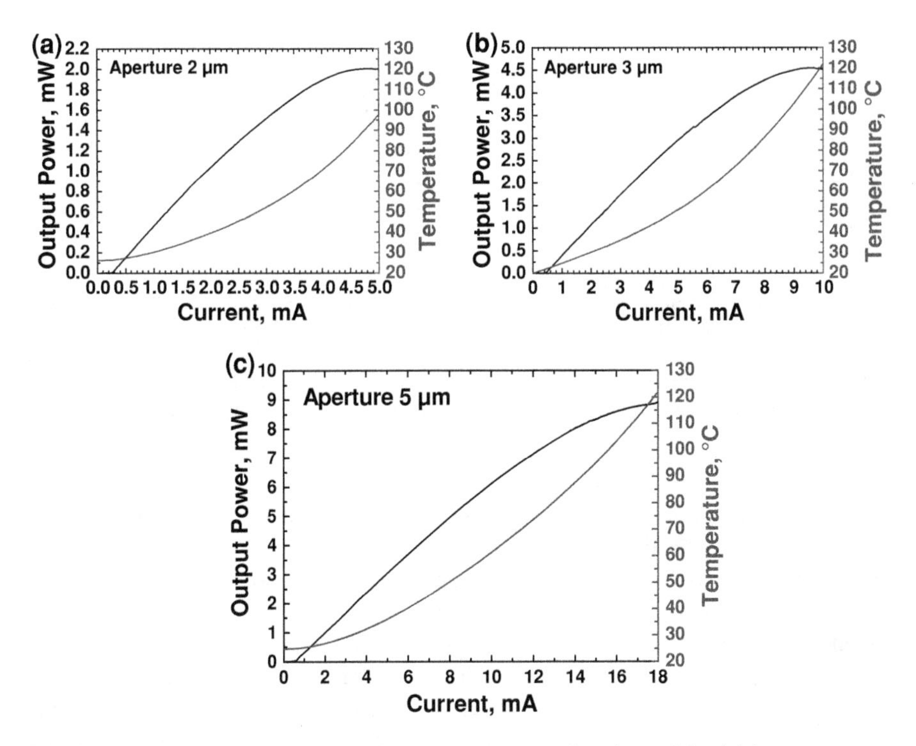

Fig. 4.7 Output power and estimated inner temperature as functions of the driving current at a fixed chuck temperature of 25°C for the 2 μm VCSEL (**a**), 3 μm VCSEL (**b**) and 5 μm VCSEL (**c**)

4.1.3 Small Signal Modulation Analysis

Small signal modulation response and scattering parameters were measured on the VCSELs described in the previous section using a network analyzer and a calibrated photodetector in the range from 50 MHz to 20 GHz. The extraction procedure of the physical parameters was similar to that described in [11] and was as follow. First, the measured optical small signal modulation response (S_{21}) was corrected for the response of the photodetector. Then, using the equivalent circuit model shown in Fig. 2.27b and described in the previous chapter, the parasitic resistances and capacitances inside of the VCSEL were extracted by fitting the circuit model to the measured microwave reflection curves (S_{11}). After that, using the values for the parasitic elements and the equivalent circuit model, the parasitic low pass curve for the electrical microwave transmission was calculated using (2.2.35–2.2.38) and subtracted from the calibrated measured optical data. Finally the resonance frequency f_R and damping factor γ were extracted using the theoretical rate equation (2.4.57), which is equivalent to (2.4.75) without the parasitic low pass. This method of data extraction has several advantages as compared to methods that fit the measured S_{21} curve data to the full equation (2.4.75), to get all three fitting parameters—resonance frequency, damping factor and parasitic cut-off frequency—only from one

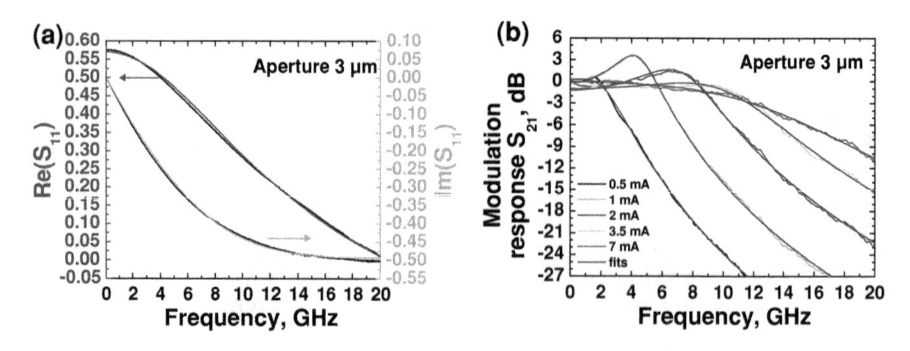

Fig. 4.8 Measured scattering parameters S_{11} at 6 mA at 25°C with corresponding fits (**a**) and measured modulation response S_{21} at different currents at 25°C with corresponding fits (**b**) for the 3 μm VCSEL

fit. First, the use of more experimental information (here the measured S_{21} and S_{11} curves) minimizes the error of the extracted parameters. The second advantage is that one is not limited to the first order low pass response of the form found in the second part of the equation (2.4.75) with the parasitic cut-off frequency f_p. Instead the calculated form of the parasitic low pass response based on the parameters extracted from the measured S_{11} data is used and this makes the fits more exact.

To illustrate the extracting procedure, real and imaginary part of the measured scattering parameters S_{11} at 6 mA and 25°C for the 3 μm VCSEL are shown in Fig. 4.8a. Corresponding fits are presented also, showing an excellent agreement with experiment. For each value of the current electrical parasitics are calculated from the parameters estimated by the fits of the corresponding S_{11} curves.

Using the calculated electrical parasitic low pass curves S_{11}, modulation response curves S_{21} for different currents can be fitted and are shown in Fig. 4.8b for the 3 μm VCSEL at 25°C. Again, all fits demonstrate an excellent agreement with experimental data. From each modulation response fit S_{21} relaxation resonance frequency f_R and damping factor γ are obtained. Together with the parasitic cut-off frequency f_p obtained from the fits of the scattering parameters S_{11}, these are the three parameters, which can be directly estimated from the measured S_{11} and S_{21} for every current value. Combining parameters obtained from the measurement data at different currents one gets access via linear fits to the *D*-factor, the *K*-factor and the damping factor offset γ_0, according to Eqs. 2.4.70 and 2.4.65. Figure 4.9 demonstrates the described relationships for the 3 μm VCSEL at 25°C. Hereby in the case of the relaxation resonance frequency only several points at lower currents should be considered, in order to exclude thermal and also gain compression effects. From the fit the *D*-factor of ~5.58 GHz/Sqrt(mA) is obtained for the 3 μm VCSEL at 25°C.

As can be seen from Fig. 4.9, linear fits match very well to the experimental data, making extraction of the parameters possible. According to the last term in (2.4.63), several points near the threshold in Fig. 4.9b have larger value of the damping factor because of the small photon density N_p, but at higher currents this term becomes negligible and all points match very well to the linear fit. Thus the

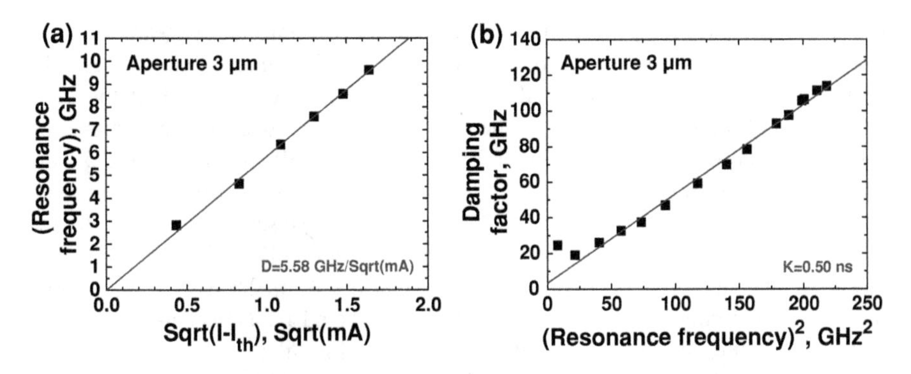

Fig. 4.9 Relaxation resonance frequency as a function of the square root of the current above threshold with extracted D-factor (**a**) and damping factor as a function of the squared relaxation resonance frequency with extracted K-factor (**b**) with corresponding linear fits for the 3 μm VCSEL at 25°C

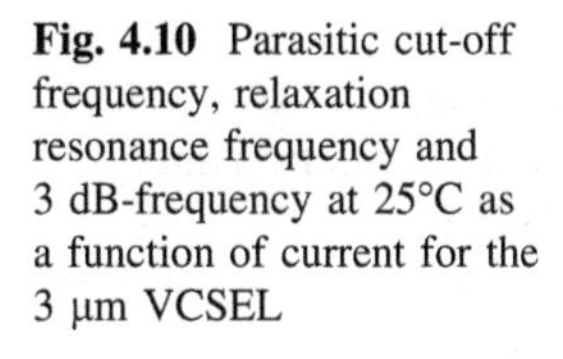
Fig. 4.10 Parasitic cut-off frequency, relaxation resonance frequency and 3 dB-frequency at 25°C as a function of current for the 3 μm VCSEL

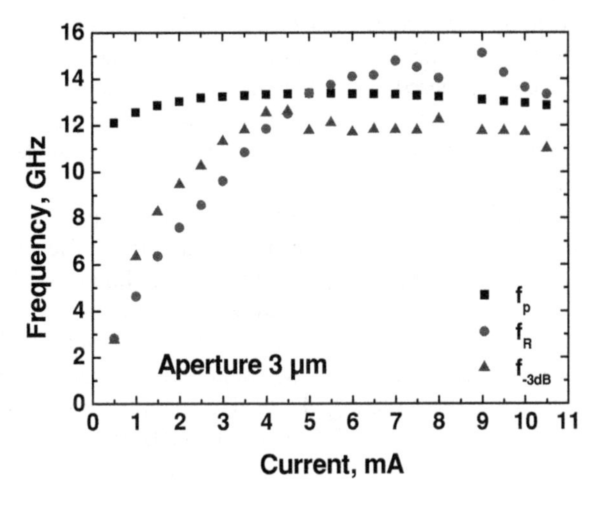

K-factor can be easily obtained and is for the 3 μm VCSEL 0.50 nm at 25°C. It corresponds to the maximum achievable bandwidth of ~17.7 GHz, if the laser would be limited only by damping. In some cases extracted values of the damping factor show a super linear dependence on the squared relaxation resonance frequency, mainly because of the thermal and gain compression effects, making extraction of the K-factor difficult. In such cases directly investigation of the form of the modulation transfer function curves at different currents and of the dependence of the relaxation resonance frequency and bandwidth on the current helps to find out factors limiting high speed operation of the VCSEL and the role of the damping in these processes.

Figure 4.10 shows three characteristic frequencies: the relaxation resonance frequency f_R, the parasitic cut-off frequency f_p and the bandwidth or the 3 dB-frequency f_{-3dB}, extracted at different currents at 25°C for the 3 μm VCSEL.

From this figure limiting factors of the high speed operation can be clearly located. While the relaxation resonance frequency achieves values larger than 15 GHz, the maximum bandwidth is ~12.5 GHz at 4.5 mA. This fact shows, that the laser speed is not crucial limited by the thermal effects, because the relaxation resonance frequency rises to higher values as the 3 dB-frequency, and in the absence of other limiting factors the maximum thermal limited bandwidth would be larger than 23 GHz, according to (2.4.76). As we have seen from Fig. 4.9b, the maximum damping limited bandwidth would be about ~17.7 GHz, which comes closer to the measured bandwidth values. Thus the 3 μm VCSEL is noticeably limited by damping. Finally, considering the extracted parasitic cut-off frequencies, which are around ~13 GHz, one can realize that electrical parasitics represent the second strong limitation. Thus the 3 μm VCSEL is limited not by the thermal effects but by a combination of damping and electrical parasitics.

For the high temperature stable VCSELs it is of a great importance to understand temperature dependence of the main laser parameters. It can be enabled by carrying out small signal modulation measurements at different temperatures, starting from the room temperature. Since the 3 μm VCSEL shows high speed behavior similar to the single-mode 2 μm VCSEL, as we will see in the following, only two devices can be considered for the temperature dependent small signal modulation analysis: the 2 μm single-mode VCSEL and the 5 μm multimode VCSEL.

The small signal modulation response of the single-mode 2 μm device at 25°C and at different currents is shown in Fig. 4.11a, together with the fits of the modulation transfer function used for the parameter extraction. Figure 4.11b shows data at 4 mA between 25 and 120°C. The maximum 3 dB-bandwidth at 25°C is 12.9 GHz, comparable to the maximum bandwidth of the 3 μm VCSEL.

The bandwidth of the 2 μm VCSEL at constant current is hardly temperature dependent and changes at 4 mA between 25 and 120°C by <2 GHz from 12.9 to 11 GHz. The reason is the weak temperature dependence of the resonance frequency, shown in Fig. 4.12a. The maximum resonance frequency changes only from ~14.5 GHz at 25°C to ~10.7 GHz at 120°C. The *D*-factor, obtained from

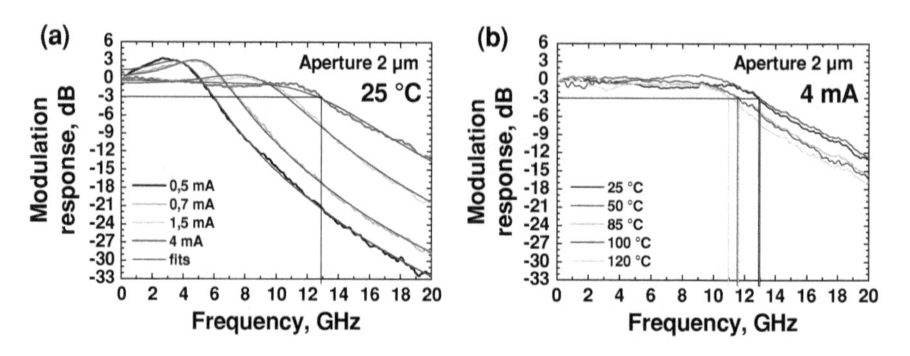

Fig. 4.11 Small signal modulation response with corresponding fits at 25°C for different currents (**a**) and small signal modulation response at a constant current of 4 mA for temperatures between 25 and 120°C (**b**) for the 2 μm VCSEL

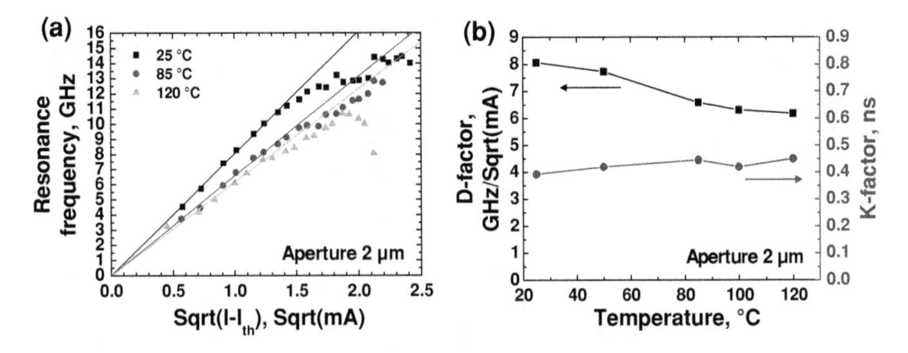

Fig. 4.12 Relaxation resonance frequency as a function of the square root of the current over threshold at 25, 85 and 120°C with corresponding fits of the linear parts (**a**) and *D*-factor and *K*-factor as a function of the temperature for the 2 μm VCSEL

the linear fit of the resonance frequency over the square root of the current above threshold for low currents, is shown as a function of the temperature in Fig. 4.12b.

The *D*-factor changes only by less than 25% from 8.1 GHz/Sqrt(mA) at 25°C to 6.1 GHz/Sqrt(mA) at 120°C. According to Eq. 2.4.70 the changes of the *D*-factor are caused by the changes of the differential gain a, mode volume V_P and internal quantum efficiency η_i. If one assumes that internal quantum efficiency and mode volume do not change with temperature noticeable, then the changes of the *D*-factor would reproduce the changes of the differential gain. In reality however, the mode volume and internal quantum efficiency are temperature dependent, but the temperature dependence of their combination together with differential gain in the presented device is weak.

Estimated *K*-factors at different temperatures from 25°C up to 120°C are shown in Fig. 4.12b as well. The *K*-factor is almost constant and is in the range ~0.40–0.45 ns. The maximum possible bandwidth with this *K*-factor is ~20 GHz. The resonance frequency in the single-mode 2 μm device reaches values larger than the 3 dB-frequency. The parasitics bandwidth is even lower than the 3 dB-frequency and is around ~9 GHz. The bandwidth is therefore limited by the combination of damping and electrical parasitics. The main source of electrical parasitics are two large capacitances: the pad capacitance C_p and the capacitance of the oxide aperture and junction C_a, which have the values of ~85 and ~200 ff, respectively, combined together with the relative large resistance of the active region R_a caused by the small aperture diameter.

The small signal modulation responses of the multimode 5 μm VCSEL at 25°C and at different currents with the corresponding fits are shown in Fig. 4.13a. The maximum 3 dB-bandwidth at 25°C is 13 GHz. Figure 4.13b shows S_{21} data at 12.5 mA and different temperatures. The bandwidth of the multimode 5 μm device is also weakly temperature dependent and reduces at constant current only from 13 GHz at 25°C to 9.9 GHz at 120°C. Figure 4.14a shows the resonance frequency as a function of the square root of the current over threshold and the corresponding linear fits for different temperatures. The maximum resonance frequency is 12 GHz at 25°C

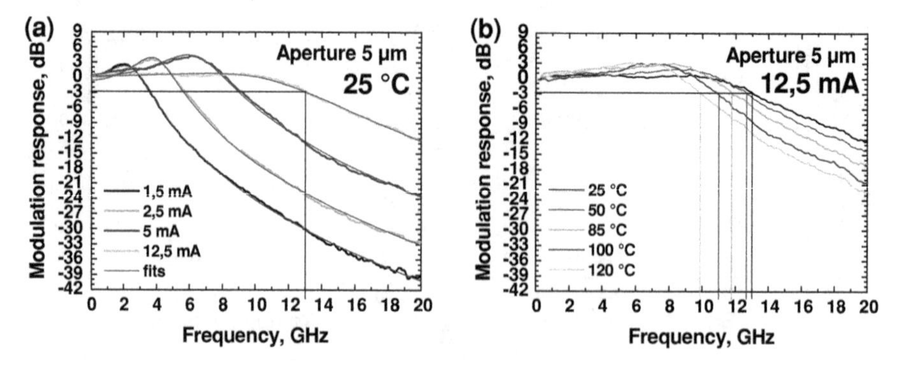

Fig. 4.13 Small signal modulation response with corresponding fits at 25°C for different currents (**a**) and small signal modulation response at a constant current of 12.5 mA for temperatures between 25 and 120°C (**b**) for the 5 μm VCSEL

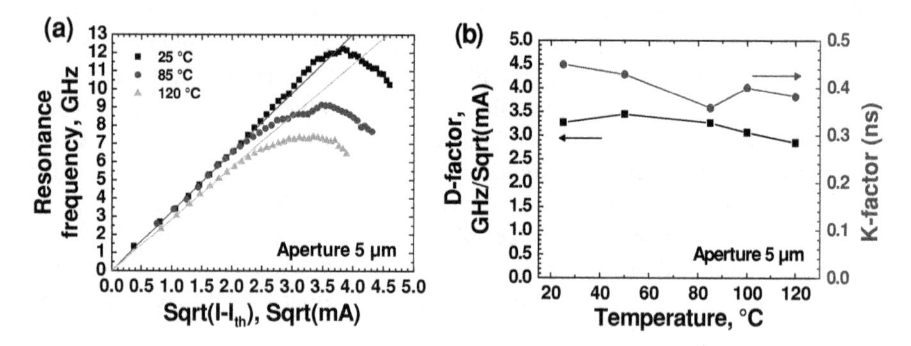

Fig. 4.14 Relaxation resonance frequency as a function of the square root of the current over threshold at 25, 85 and 120°C with corresponding fits of the linear parts (**a**) and *D*-factor and *K*-factor as a function of the temperature for the 5 μm VCSEL

and 7.3 GHz at 120°C. These values are smaller as compared to the resonance frequencies of the single-mode 2 μm VCSEL and of the 3 μm VCSEL. The reason is the smaller *D*-factor due to the larger mode volume V_P of the multimode 5 μm VCSEL as compared to the single-mode 2 μm VCSEL and also to the 3 μm VCSEL. The *D*-factor of the multimode 5 μm VCSEL is shown in Fig. 4.14b. It changes from 3.3 GHz/Sqrt(mA) at 25°C to 2.9 GHz/Sqrt(mA) at 120°C and is hardly temperature dependent. The small *D*-factor is the reason for the fact that the multimode 5 μm VCSEL can not reach a high resonance frequency before thermal roll over. At 25°C chuck temperature the temperatures inside the cavities at the beginning of the thermal roll over are for both 2 μm VCSEL and 5 μm VCSEL comparable, as we have seen previously. But because of the larger *D*-factor the single-mode 2 μm VCSEL reaches larger resonance frequencies as compared to the multimode 5 μm laser.

The *K*-factor of the multimode device is shown in Fig. 4.14b too and is also weakly temperature dependent. Its value is about ~0.40 ns, which is similar to those of the single-mode 2 μm VCSEL.

In the case of the 5 μm VCSEL the relaxation resonance frequencies are smaller than the 3 dB-bandwidth. Thus the multimode laser is limited noticeable by the smaller D-factors or in other words by the smaller relaxation resonance frequencies and thus by the thermal effects. The bandwidth of the electrical parasitics is ~14 GHz and additionally restricts the 3 dB-bandwidth. Parasitic capacitances are ~65 fF for the C_p and ~290 fF for the C_a.

Performed small signal measurements have shown excellent temperature stability of the investigated VCSELs, which results mainly from the optimized active region and cavity. Indicated presence of confinement centers in the active region would result in a decreased carrier in-plane mobility improving the high-temperature performance. In small aperture VCSELs with conventional QWs a significant portion of carriers are lost at high temperatures due to carrier reemission into the separate confinement heterostructure region and lateral carrier diffusion out of the active area. Thus a reduced in-plane mobility of the carrier in the active region improves the temperature stability of small aperture devices. The presence of the gain offset balances the decreasing peak differential gain with temperature, increasing its temperature stability further [12]. These optimization steps have resulted in measured bandwidths around ~13 GHz at 25°C and ~10 GHz at 120°C, which should enable device operation at 20 Gbit/s at all temperatures from 25 up to 120°C.

Additionally, small signal analysis has enabled to find out factors limiting high speed operation of the lasers. It has been recognized, that all investigated VCSELs are noticeable limited by electrical parasitics. Thus to increase the bandwidth, reduction of the parasitic capacitances is indispensable. Damping represents the second limitation, which is weaker and limits the bandwidth to values of ~18–20 GHz. Thermal heating limits larger VCSELs and can be overcome by increasing the D-factor, which means by optimizing the active region and cavity, and also by reducing thermal resistance of lasers. It will open the way to reach larger relaxation resonance frequencies in VCSELs with larger aperture diameters, reducing current density and improving device reliability.

4.1.4 Large Signal Modulation Characteristics

An application oriented proof of the high speed properties of a VCSEL are large signal modulation measurements or the so called optical eye diagrams. Thereby the eye form, the eye extinction ratio (ER) and the eye signal-to-noise ratio (SNR) are among other important characteristics, which give access to the laser behavior during the data transmission process at a defined bit rate.

Optical eye diagrams for the three VCSELs: the 2 μm single-mode VCSEL, the 3 μm VCSEL and the 5 μm multimode VCSEL, measured at 25°C using a 2^7-1 pseudo-random bit sequence (PRBS) at 20 Gbit/s in a back-to-back (BTB) configuration (~3 m fiber) with an SHF 12100 B bit pattern generator and a New Focus 1554-B-50 photoreceiver are shown in Fig. 4.15.

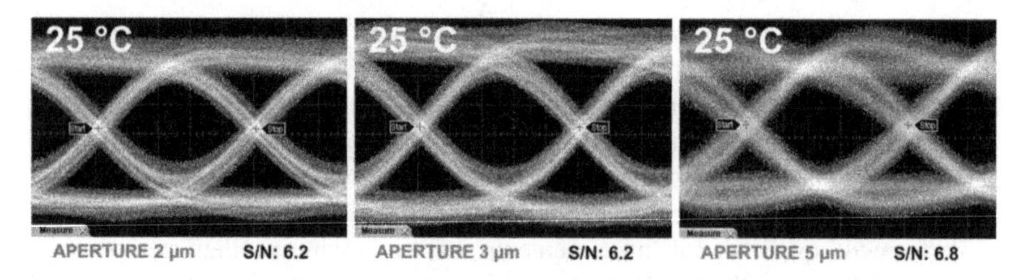

Fig. 4.15 20 Gbit/s 2^7-1 PRBS back-to-back eye diagrams and signal-to-noise ratios at 25°C for the 2 μm VCSEL (*left*), 3 μm VCSEL (*middle*) and 5 μm VCSEL (*right*)

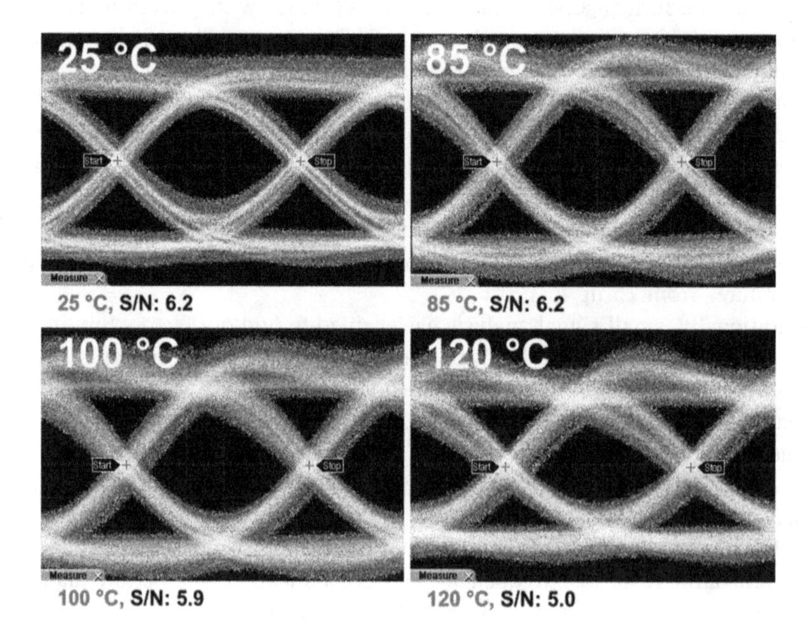

Fig. 4.16 20 Gbit/s 2^7-1 PRBS back-to-back eye diagrams and signal-to-noise ratios between 25 and 120°C for the constant bias current of 4.3 mA and modulation voltage of 1.22 V for the 2 μm single-mode VCSEL

The eyes are clearly open for all three VCSELs and have SNR larger than 6. Extinction ratio was above 4 dB for all measurements. The rise and fall times of the traces are in parts limited by the photoreceiver bandwidth of ~12 GHz. For the smaller two VCSELs—the 2 μm VCSEL and the 3 μm VCSEL, the eye form is symmetric and no noticeable overshoot is observed. The form of the multimode 5 μm VCSEL becomes asymmetric, with several traces and an overshoot. While at temperatures up to 100°C optical eyes from all three devices remain comparable, at higher temperatures the two smaller VCSELs demonstrate better eye form and SNR, so that in the following these two devices would be investigated closer.

In Fig. 4.16 optical eye diagrams at 20 Gbit/s for temperatures between 25 and 120°C for the single-mode 2 μm VCSEL are shown, measured in the same configuration as eye diagrams in Fig. 4.15. The measured signal-to-noise ratios are also given. The bias current and peak-to-peak modulation voltage V_{p-p} were kept constant at 4.3 mA and 1.22 V for all measurements, respectively. This gives the possibility to drive the VCSEL in the possible future application without additionally electronics, making its installation and operation cheaper. The eyes are clearly open and error free at all temperatures from 25 up to 120°C. The SNR is 6.2 at 25°C and 5.0 at 120°C; extinction ratio was above 4 dB for all measured eye diagrams.

The 3 μm VCSEL has demonstrated even better results and has the best large signal modulation characteristics compared to all investigated 980 nm SML-VCSELs. Optical eye diagrams of the 3 μm VCSEL were measured using the same measurement configuration as described above. Again, as in the case of the 2 μm VCSEL, for all measurements with the 3 μm VCSEL the bias current and modulation voltage V_{p-p} were kept constant at 6 mA and 1.2 V, respectively. Figure 4.17 shows the results for 20 Gbit/s for temperatures between 0 and 120°C. The eyes are clearly open over the whole temperature range up to 120°C. The SNR is 6.0 at 0°C, 6.2 at 25°C and 5.3 at 120°C. The extinction ratio was above 4 dB for all temperatures. The 3 μm 980 nm SML-VCSEL is the first of any VCSELs at any wavelength showing a 20 Gbit/s open eye at 120°C, to the best of our knowledge.

As one can see, all investigated 980 nm SML-VCSELs are able to operate at bit rates as high as 20 Gbit/s at room temperature. Because of their optimized structure the operation temperature can be increased up to 120°C for the smaller VCSELs without adjustment of the driving conditions, leaving the eye quality hardly affected. The eyes remain clearly open, ensuring a proper VCSEL operation during the data transmission process at 20 Gbit/s at temperatures from 0°C in the case of the 3 μm VCSEL and from 25°C for the 2 μm VCSEL up to 120°C for both devices.

Thus, large signal modulation experiments have demonstrated the suitability of the 980 nm SML-VCSELs to transmit data at 20 Gbit/s in the temperature range from 0 to 120°C and confirmed high temperature stability of the investigated lasers.

4.1.5 Summary of the 980 nm SML-VCSEL Results

Optimization of the cavity region and both mirrors for high thermal conductivity, low electrical resistance and low optical losses, together with the usage of the highly strained InGaAs layers grown in the submonolayer growth mode in the active region, enhancing its the thermal stability, have led to realization of high temperature stable 980 nm VCSELs operating at 20 Gbit/s at temperatures from 0 up to 120°C. These were first VCSELs showing an open eye at 120°C at 20 Gbit/s. Both CW and high frequency laser characteristics have demonstrated very stable behavior within a wide temperature range up to 120°C.

Small signal modulation experiments were carried out and provided access to the key physical parameters of investigated 980 nm SML-VCSELs, among other

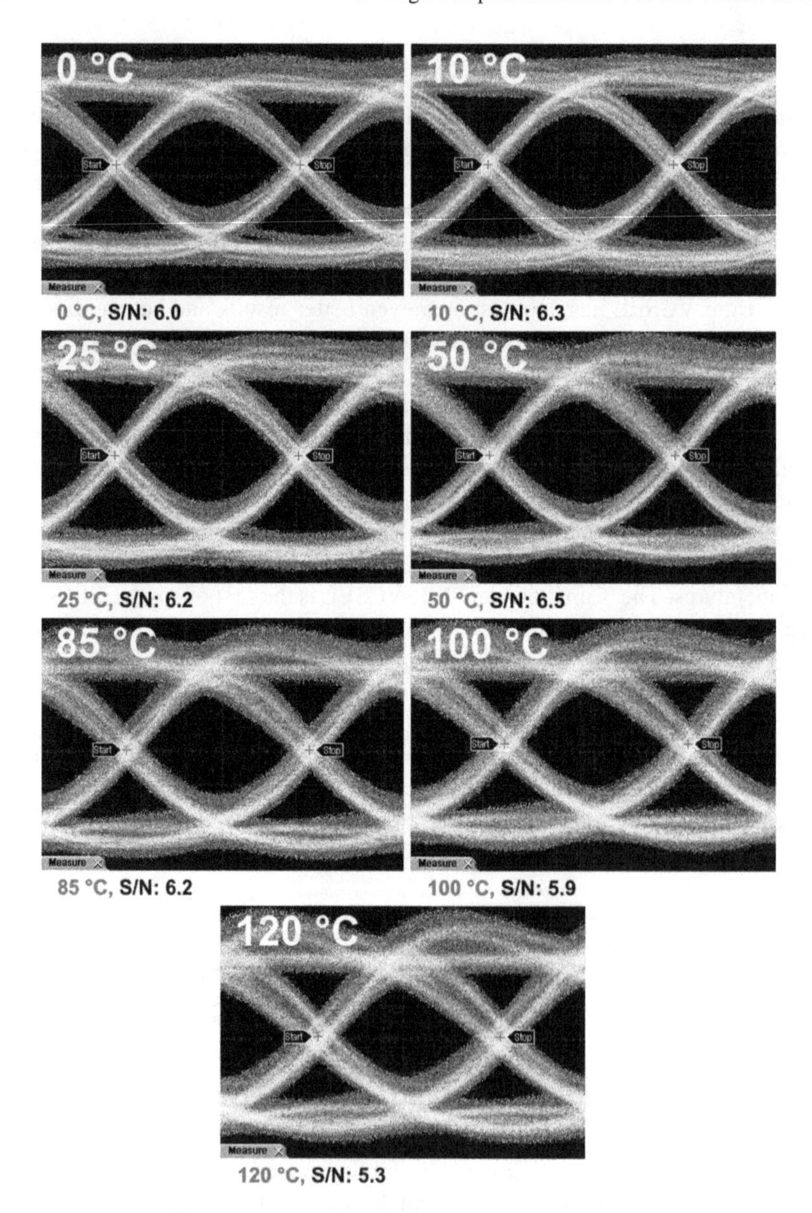

Fig. 4.17 20 Gbit/s 2^7-1 PRBS back-to-back eye diagrams and signal-to-noise ratios between 0 and 120°C for the constant bias current of 6 mA and modulation voltage of 1.2 V for the 3 μm VCSEL

to relaxation resonance frequency, 3 dB-frequency, parasitic cut-off frequency, damping factor, *D*-factor and *K*-factor. By analyzing these parameters and their dependences on current and temperature, factors limiting the high speed laser operation have been evaluated. It has turned out that the 980 nm SML-VCSELs with smaller oxide aperture diameters in the range of 2–3 μm are limited mainly by electrical parasitics and also by damping. Lasers with larger oxide aperture diameters were limited by a combination of electrical parasitics and thermal effects. Also damping plays some role in the limitation at larger apertures.

Consequently, to improve high speed properties of the VCSELs further, one needs to overcome these limitations. First of all electrical parasitics should be suppressed, which can be achieved by applying multiple oxide apertures and by reducing the laser mesa diameter. While in the 980 nm SML-VCSELs only one oxide aperture was applied and laser mesa diameters in the range of ~34–36 μm were used, one can apply in the next generation devices with two apertures and mesa diameters of ~25–30 μm, which should reduce parasitic capacitances noticeable. To overcome the damping and thermal limitations further optimization of active region and cavity should be applied. In the next Sect. 980 nm QW-VCSELs will be presented, where improvements of both epitaxial structure and device fabrication have led to large signal operation at bit rates of 25 Gbit/s and higher.

4.2 VCSELs with the QW Active Region

From the measurement results of the 980 nm SML-VCSELs, presented in the previous section, physical processes limiting high speed device operation have been identified. The strongest limitation was caused by electrical parasitics. Additionally, damping played a noticeable role, while thermal processes had a weaker effect on high speed laser properties. To overcome these limitations the second generation of 980 nm VCSELs—the 980 nm QW-VCSELs—was designed, grown, fabricated and analyzed. The main difference to the 980 nm SML-VCSELs grown by MBE, which were presented in the previous section, was the growth of the 980 nm QW-VCSEL structures by the MOCVD technique, moving these devices one step closer to a possible large scale production, and also introduction of the compressively strained InGaAs QWs as active region instead of the InGaAs layers grown in the submonolayer growth mode, making the 980 nm QW-VCSELs more insensitive and robust to possible variations of the growth process. Further optimization of the VCSEL structure enabled to increase the laser bandwidth by more than 20% and achieve bit rate of 25 Gbit/s at 85°C.

4.2.1 Device Structure

The growth of the 980 nm QW-VCSELs was performed by MOCVD on the n-doped GaAs (100) substrates. Laser active region consisted of five compressively strained 4.2 nm thick $In_{0.21}Ga_{0.79}As$ QWs separated and surrounded by 6 nm thick $GaAs_{0.88}P_{0.12}$ tensile strained compensation layers to ensure high quality defect-free crystal growth of the VCSEL structure. By introduction of the compressive strain to the active QWs their transparency carrier density is reduced and their differential gain is enhanced [13]. Both effects improve high speed properties of the lasers. The QWs together with the strain compensation layers were surrounded by the two 20 nm thick $Al_{0.35}Ga_{0.65}As$ barrier layers to improve carrier localization in the active region.

In order to overcome limitations for high speed operation, which have been identified in the previous section for the 980 nm SML-VCSELs, also cavity region of the 980 nm QW-VCSELs has been optimized. To reduce electrical parasitics two 30 nm thick $Al_{0.98}Ga_{0.02}As$ aperture layers have been introduced, instead of one AlAs aperture layer in the 980 nm SML-VCSELs. The first aperture was in the first top DBR pair closest to the cavity region, and the second aperture was inside the cavity just below the DBR. This should reduce the capacitance of the oxide layers and thus increase the cut-off frequency of electrical parasitics. The 2% of Ga in the aperture layers help to reduce mechanical strain introduced during the selective wet oxidation process. The cavity length was 3/2 λ, since it is a good compromise between a short cavity and a proper distances between active layers and aperture region. Similar to the 980 nm SML-VCSELs, the cavity was red shifted from the peak gain at 25°C by nominally 15 nm in order to improve the temperature stability of the lasers. To overcome the damping limit doping profiles in the cavity region and also in the mirrors have been optimized further, in order to reduce optical losses while leaving electrical resistance acceptable low.

Doped $Al_{0.90}Ga_{0.10}As/Al_{12}Ga_{0.88}As$ DBR mirrors were applied for both top and bottom mirror. The top p-doped DBR consisted of 23.5 mirror pairs, while the bottom n-doped DBR contained 37.5 mirror pairs. Linear gradings with 20 nm thickness were applied to reduce electrical resistance of the mirrors.

With the same fabrication process, as was applied for the 980 SML-VCSELs, high speed 980 nm QW-VCSELs with co-planar top contacts and different aperture diameters were manufactured. One important difference was in the selective wet oxidation time, resulting in devices with smaller laser mesa diameters at the same oxide aperture diameters, as compared to the 980 nm SML-VCSELs. While in the case of the 980 nm SML-VCSELs, described in the previous section, oxide aperture diameters of $\sim$5 μm corresponded to laser mesa diameters of $\sim$35 μm, in the case of the 980 nm QW-VCSELs the same aperture diameters corresponded to laser mesa diameters of $\sim$30 μm, which is 5 μm smaller, additionally reducing parasitic capacitances. Fields, each containing ten devices with different mesa and thus aperture diameters in steps of nominally 1 μm, were accessible for measurements after the high yield fabrication process, enabling detailed characterization of laser properties at different oxide aperture diameters.

4.2.2 Static Characteristics

Fields containing ten 980 nm QW-VCSELs with aperture diameters from 4 up to 13 μm in steps of 1 μm were investigated. Corresponding L–I and U–I

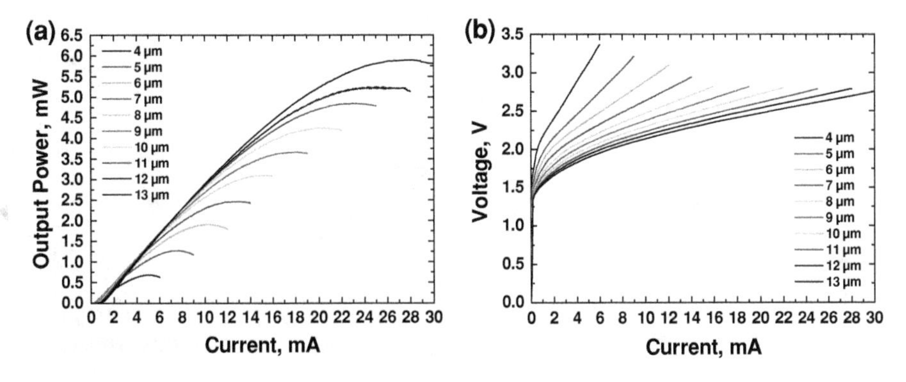

Fig. 4.18 L–I characteristics (a) and U–I characteristics (b) of a VCSEL array with aperture diameters from 4 to 13 µm in steps of 1 µm at room temperature

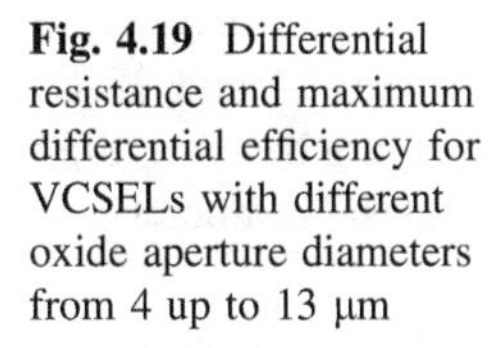
Fig. 4.19 Differential resistance and maximum differential efficiency for VCSELs with different oxide aperture diameters from 4 up to 13 µm

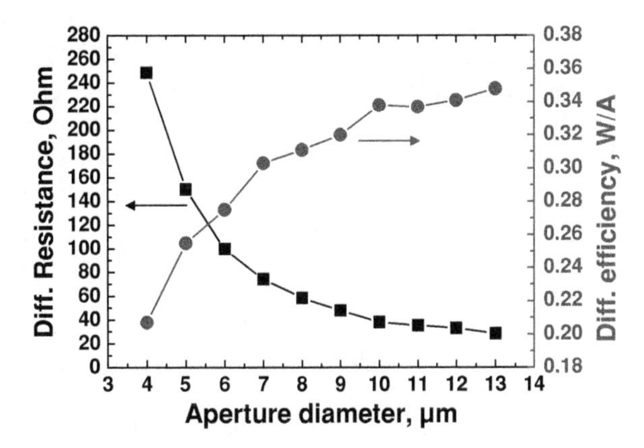

characteristics for a measured field at room temperature are shown in Fig. 4.18. Maximum output powers from ~0.7 mW for the 4 µm aperture device up to ~5.9 mW for the 13 µm aperture device were measured. Extracted values of the differential resistances and maximum differential efficiencies for the measured VCSELs are shown in Fig. 4.19. According to the increase of the aperture diameter differential resistance decrease from ~250 Ω for the 4 µm aperture VCSEL down to the ~30 Ω for the 13 µm aperture device. The maximum differential efficiency increases with the aperture diameter from ~0.21 W/A for the 4 µm aperture up to ~0.35 W/A for the 13 µm aperture. The increase of the differential efficiency with the oxide aperture diameter could be explained according to (2.4.19) and (2.4.20) by improved current injection efficiency η_i at larger apertures and reduced effect of the scattering losses, which are included into the internal cavity losses α_i. By analyzing the CW data of the investigated VCSELs with different oxide aperture diameters can be expected, that devices with aperture diameters in the range of 7–10 µm should be mostly suitable for high speed operation, because of their sufficient large maximum output powers, which are

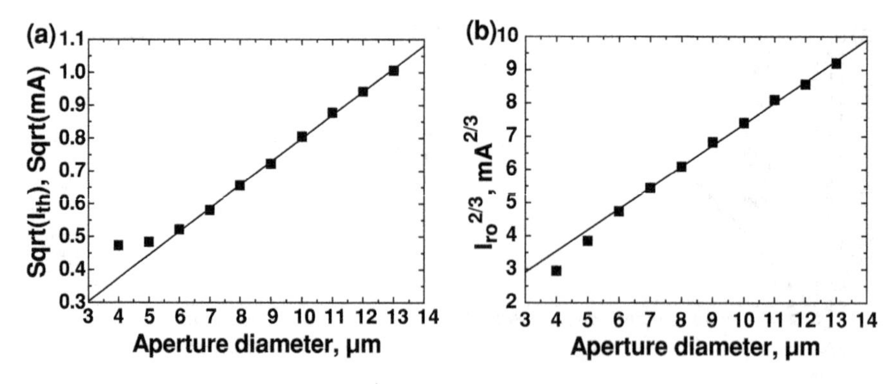

Fig. 4.20 Square root of the threshold current (**a**) and the roll-over current to the 2/3 power (**b**) for VCSELs with different oxide aperture diameters with corresponding linear fits

between 2.5 mW for the 7 μm aperture device and 4.3 mW for the 10 μm aperture VCSEL, sufficient low differential resistances, which are between 74 and 38 Ω, and reasonable values of the maximum differential efficiencies, lying between 0.30 and 0.34 W/A. Additionally, these devices have nominal mesa diameters of 30–33 μm, resulting in lower parasitic capacitances.

By plotting the square root of the threshold current as a function of the oxide aperture diameter one can prove, whether all investigated VCSELs have similar threshold current density. In fact, as can be seen from the Fig. 4.20a, measurement data of the eight from the ten investigated VCSELs match to the linear fit. The deviation of the measured threshold currents for the two lasers with smaller aperture diameters could be explained by the edge effects, caused by the inhomogeneous current distribution.

According to (2.3.17) one can plot the roll-over current to the power of 2/3 as a function of the aperture diameter, as shown in Fig. 4.20b. Again, all measured devices, excepting the two VCSELs with the smallest aperture diameters, match very well to the linear fit. Both diagrams in Fig. 4.20 confirm the high quality of the growth and fabrication processes, resulting in defined and predictable laser parameters, including among other mesa diameters, oxide aperture diameters, threshold currents, etc.

Since the most suitable for high speed operation VCSELs have aperture diameters in the range from 7 to 10 μm, temperature dependent CW characteristics were measured for the 7 μm aperture and for the 10 μm aperture devices and are shown together with the extracted temperature dependent values of the maximum differential efficiency and threshold current in Figs. 4.21 and 4.22, respectively.

As follows from Fig. 4.21a, the maximum output power is very temperature stable and decreases from ~2.5 mW at 20°C to ~1.5 mW at 85°C, which is ~60% of the value at 20°C. This change of the output power with temperature is comparable to the 980 nm SML-VCSELs investigated in the previous section, where the maximum output power at 85°C had values of ~58–64% of the value at

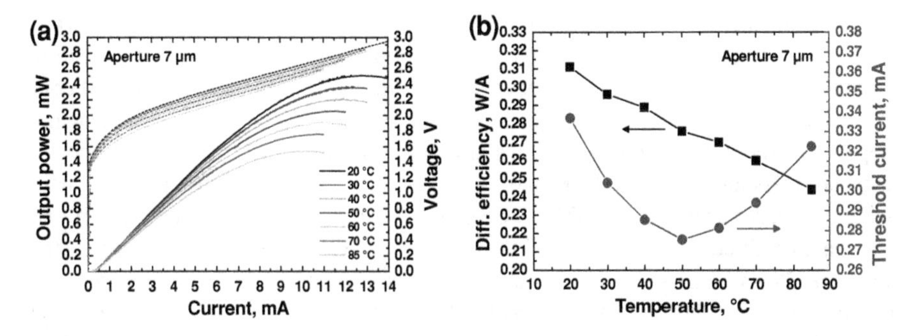

Fig. 4.21 L–U–I characteristics (**a**) and extracted values of the maximum differential efficiency and threshold current (**b**) at different temperatures between 20 and 85°C for the VCSEL with 7 μm aperture

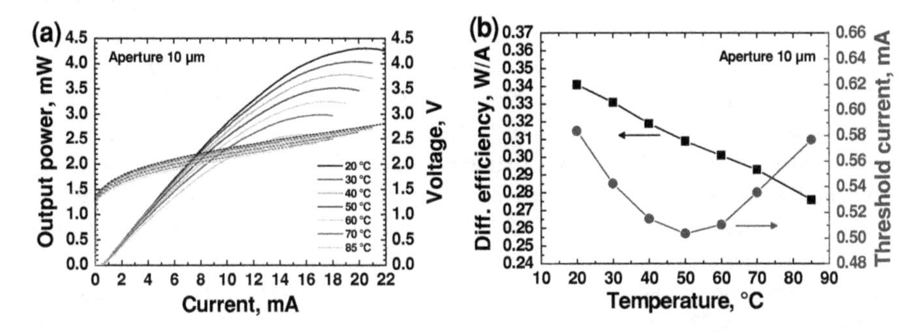

Fig. 4.22 L–I characteristics (**a**) and extracted values of the maximum differential efficiency and threshold current (**b**) at different temperatures between 20 and 85°C for the VCSEL with 10 μm aperture

25°C for different aperture diameters. The threshold voltages are in the range of ~1.5 V for all measured temperatures, comparable to the 980 nm SML-VCSELs. The maximum differential efficiency shown in Fig. 4.21b is also hardly temperature dependent and decreases from ~0.31 W/A at 20°C only to ~0.24 W/A at 85°C. The changes of ~22% in the case of the 980 nm QW-VCSELs are comparable to the changes of ~19% in the case of the 980 nm SML-VCSELs. One of the main effects contributing to the high temperature stability of the investigated QW-VCSELs is the 15 nm detuning between the gain peak and the cavity dip at room temperature, which is nominally identical to the value applied in the SML-VCSELs described in the previous section. The effect of the detuning can be directly observed by taking a look at the threshold current as a function of temperature, shown in Fig. 4.21b. A minimum around 50°C can be clearly seen, resulting in the overall changes of the threshold current in the whole temperature range from 20 up to 85°C of only ~61 μA, which corresponds to only ~19% of the maximum value. The threshold current is ~337 μA at 20°C, decreases to ~275 μA at 50°C and then increases to ~323 μA at 85°C. The high temperature stability of the threshold current, and thus of the threshold carrier density, is one of the most important preconditions for the high temperature stability of the high speed laser properties, as follows from the rate equation theory described previously, where important laser parameters, e.g. differential gain, depend on the carrier density. From the measured temperature dependent spectra thermal resistance of ~3.23 K/mW was obtained for the 7 μm QW-VCSEL, which is larger than for the SML-VCSELs described in the previous section and can be explained by the usage of the ternary alloys in the DBR mirrors in the 980 nm QW-VCSELs, leading to a lower thermal conductivity of these devices compared to the 980 nm SML-VCSELs. Nevertheless, the measured value of the thermal conductivity is sufficient to ensure acceptable temperatures of the active region and confirms the suitability of the MOCVD grown ternary mirrors for high temperature stable laser operation.

In Fig. 4.22a, measured L–U–I characteristics for the 10 μm aperture VCSEL for temperature between 20 and 85°C are presented. The maximum output power changes only from ~4.3 mW at 20°C to ~2.6 mW at 85°C. The change is only 40% of the value at 20°C, which is practically identical to the 7 μm aperture QW-VCSEL and to the SML-VCSELs investigated previously. The threshold voltage is also in the range of ~1.5 V, similar to the 7 μm aperture QW-VCSEL. Figure 4.22b shows extracted values for the maximum differential efficiency and threshold current as a function of temperature. The differential efficiency decreases only from ~0.34 W/A at 20°C to ~0.27 W/A at 85°C and the relative change is comparable to the 7 μm aperture QW-VCSEL and to the SML-VCSELs presented in the previous section. The measured threshold current shown in Fig. 4.22b exhibits a clear minimum around 50°C, resulting from the gain–cavity detuning. It has the value of ~584 μA at 20°C, reduces to ~504 μA at 50°C and then increases up to ~577 μA at 85°C. The overall relative changes in the whole temperature range are even smaller as compared to the 7 μm aperture QW-VCSEL and are only ~14%. The measured thermal resistance is ~2.24 K/mW, which is logically lower as that of the 7 μm laser, because of the larger oxide aperture diameter of the 10 μm aperture VCSEL, according to (2.3.10). In fact, dividing the value of 3.23 K/mW for the 7 μm aperture VCSEL by the value of 2.24 K/mW for the 10 μm aperture device, results in a factor of ~1.44, practically identical to the reverse proportion of the oxide aperture diameters 10–7, which is equal to ~1.43, fully confirming the theoretical prediction.

Finally, Fig. 4.23 shows emission spectra of the 7 μm aperture VCSEL and of the 10 μm aperture VCSEL at 20°C. Both devices operate at wavelengths near 980 nm and both spectra are multimode because of the relatively large aperture diameters.

Measured CW characteristics confirmed high quality growth and fabrication processes, leading to predictable laser characteristics for different oxide aperture diameters. Temperature dependent measurements demonstrated high temperature stability of the static characteristics of the investigated 980 nm QW-VCSELs, which is decisive for high temperature stable high speed device operation.

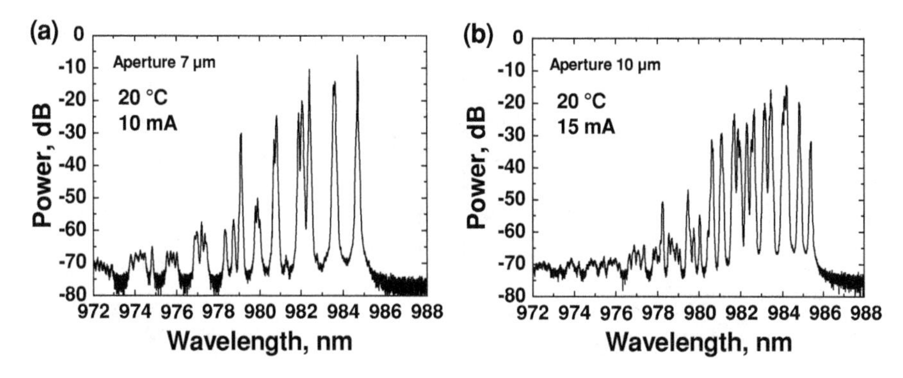

Fig. 4.23 Emission spectrum of the 7 μm aperture VCSEL at 10 mA (**a**) and of the 10 μm aperture VCSEL at 15 mA (**b**) at 20°C

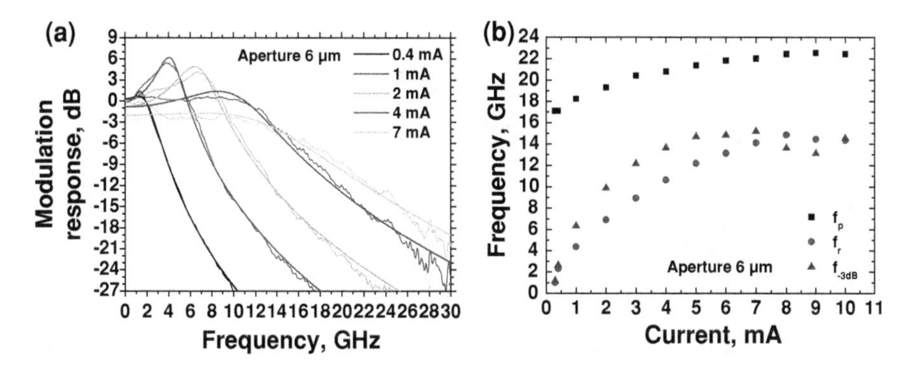

Fig. 4.24 Small signal modulation response with corresponding fits at room temperature for different driving currents (**a**) and extracted values of the parasitic cut-off frequency (*rectangles*), relaxation resonance frequency (*circles*) and 3 dB-frequency (*triangles*) as a function of the current (**b**) for the VCSEL with 6 μm aperture diameter

4.2.3 Small Signal Modulation Analysis

To investigate high speed properties of the fabricated 980 nm QW-VCSELs, small signal modulation measurements were carried out and corresponding physical parameters were extracted. The extraction procedure was the same as described earlier. Complete VCSEL field containing ten devices was characterized. Since physical properties depend on oxide aperture diameter, important laser parameters, among other parasitic cut-off frequency, relaxation resonance frequency, 3 dB-frequency, damping, *D*-factor and *K*-factor, were extracted for each of the measured VCSELs with different aperture diameters from 4 up to 13 μm.

In Figs. 4.24, 4.25 and 4.26 measured small signal modulation curves with corresponding fits and also extracted values for the parasitic cut-off frequency, relaxation resonance frequency and 3 dB-frequency are shown for three

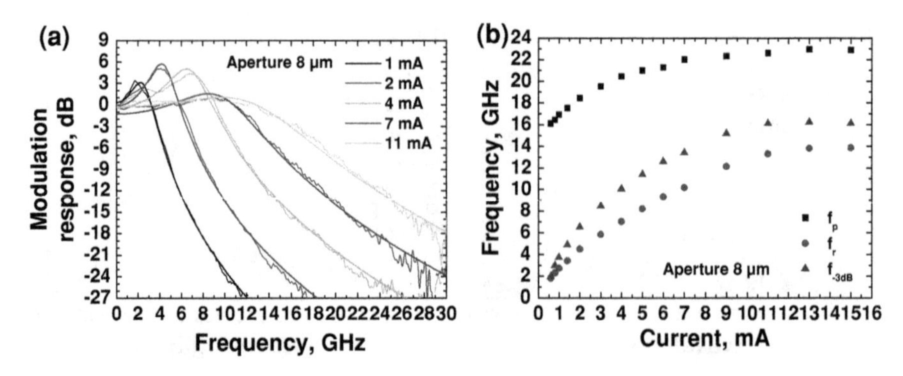

Fig. 4.25 Small signal modulation response with corresponding fits at room temperature for different driving currents (**a**) and extracted values of the parasitic cut-off frequency (*rectangles*), relaxation resonance frequency (*circles*) and 3 dB-frequency (*triangles*) as a function of the current (**b**) for the VCSEL with 8 μm aperture diameter

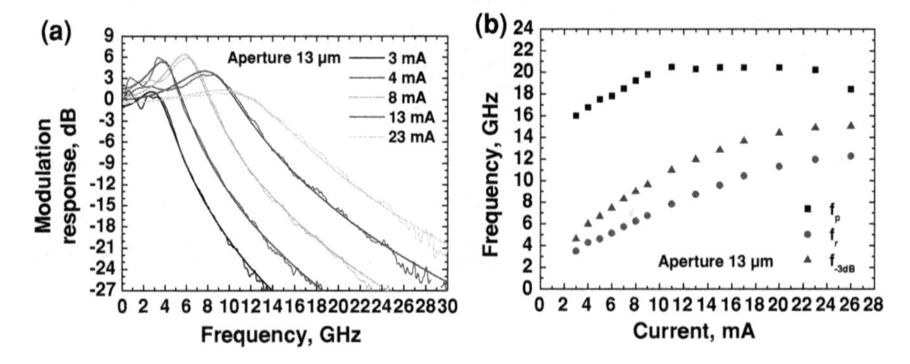

Fig. 4.26 Small signal modulation response with corresponding fits at room temperature for different driving currents (**a**) and extracted values of the parasitic cut-off frequency (*rectangles*), relaxation resonance frequency (*circles*) and 3 dB-frequency (*triangles*) as a function of the current (**b**) for the VCSEL with 13 μm aperture diameter

characteristic VCSELs with aperture diameters of 6, 8 and 13 μm. All fits reproduce experimental results very well.

All three VCSELs demonstrate similar maximum bandwidth in the range of ~15–16 GHz. The effect of electrical parasitics on high speed properties is in all three VCSELs not dominant, because of the relatively large parasitic cut-off frequency of larger than 20 GHz for all investigated devices. Although all three VCSELs have similar maximum bandwidth, each of them shows different high speed behavior, as can be obtained from the Figs. 4.24, 4.25 and 4.26. While relaxation resonance frequency of the 6 μm VCSEL reaches values comparable to or even higher than the 3 dB-frequency (Fig. 4.24), the relaxation resonance frequency of the 8 μm VCSEL is smaller than the 3 dB-frequency (Fig. 4.25) and for the 13 μm device this difference is even larger (Fig. 4.26).

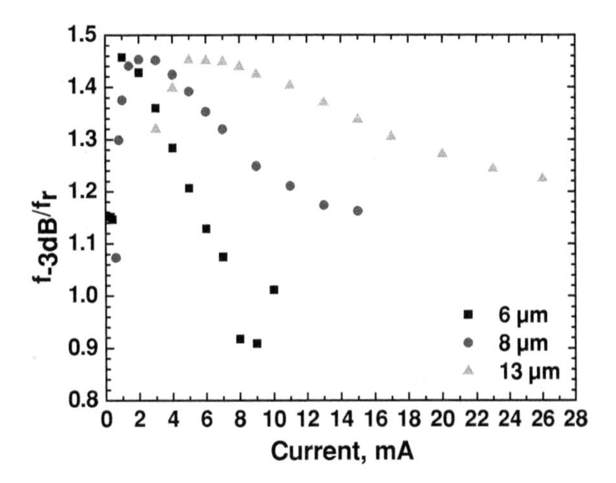

Fig. 4.27 Ration of the 3 dB-frequency to the relaxation resonance frequency as a function of current for three VCSELs with different aperture diameters of 6, 8 and 13 μm

This ratio of the 3 dB-frequency to the relaxation resonance frequency as a function of current for the three VCSELs is shown in Fig. 4.27. As can be seen, at low currents this ratio reaches for all three VCSELs values larger than 1.45, which is close to the ideal value of ~1.55, obtained for the case of absent damping and electrical parasitics (2.4.76). At larger currents the relaxation resonance frequency increases, and the ratio of the 3 dB-frequency to the relaxation resonance frequency begins to decrease, since damping starts to play a noticeable role at larger relaxation resonance frequencies and thus currents, according to (2.4.65). Because relaxation resonance frequency saturates for devices with larger aperture diameters at lower values, this ratio remains for such lasers larger, as can be obtained from Fig. 4.27. If a VCSEL would be limited mostly by thermal heating, the ratio would grow up at lower currents and then remain approximately constant. Since it is not the case for the VCSELs investigated here, one can conclude that the lasers are not limited by the thermal effects, although they play larger role for devices with larger aperture diameters, because the ratio of the 3 dB-frequency to the relaxation resonance frequency is for these devices larger as for lasers with smaller aperture diameters. Since electrical parasitics do not limit the VCSEL speed decisively, it can be concluded that damping plays a major role in limiting the VCSEL speed.

The major role of the damping can be confirmed by extracting the K-factor. In Fig. 4.28a extracted values of the D-factor and the K-factor are presented for ten investigated VCSELs with different aperture diameters from 4 up to 13 μm.

As can be seen, values between ~0.53 and ~0.37 ns for the K-factor were obtained, which corresponds to the maximum damping limited bandwidth of ~17–24 GHz, according to (2.4.77). Together with a weaker effect of electrical parasitics this leads to the measured bandwidths around ~15–16 GHz. Maximum parasitic cut-off frequencies, relaxation resonance frequencies and 3 dB-frequencies for the ten measured VCSELs with different aperture diameters are shown in Fig. 4.29, demonstrating again the smaller maximum relaxation resonance

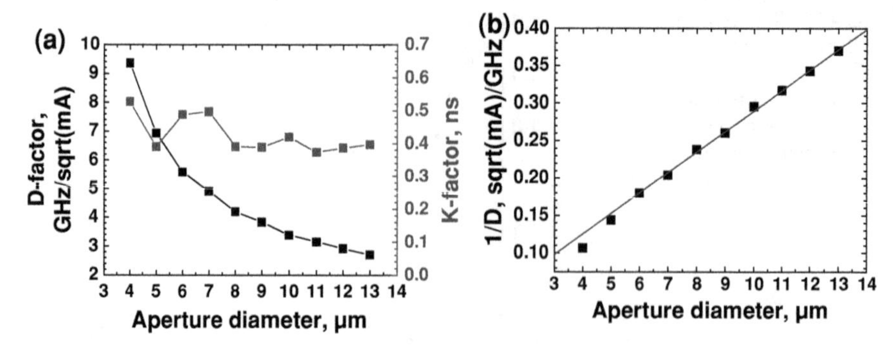

Fig. 4.28 Extracted *D*-factor and *K*-factor as a function of the oxide aperture diameter at room temperature (**a**) and inverse *D*-factor (1/D) for different aperture diameters with the corresponding linear fit (**b**)

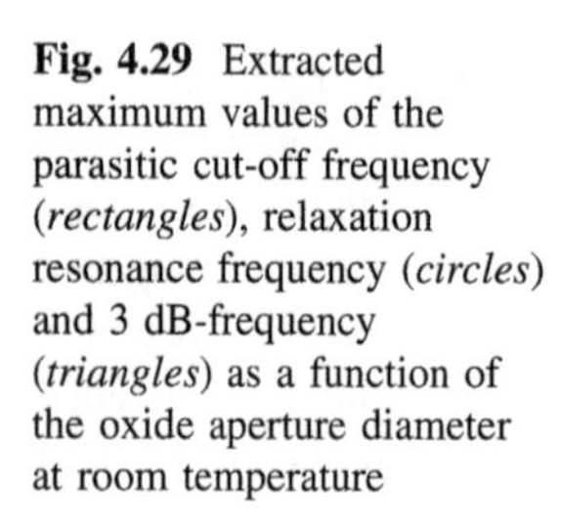

Fig. 4.29 Extracted maximum values of the parasitic cut-off frequency (*rectangles*), relaxation resonance frequency (*circles*) and 3 dB-frequency (*triangles*) as a function of the oxide aperture diameter at room temperature

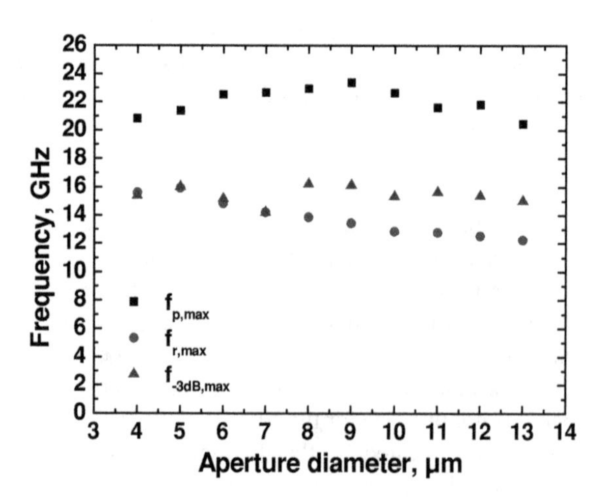

frequencies for lasers with larger apertures, while the maximum 3 dB-frequency remains comparable for all aperture diameters. The maximum parasitic cut-off frequency has a maximum at aperture diameters around ~9 µm and reaches there values larger than 23 GHz. This behavior can be explained by the trade-off between decreasing differential resistances for devices with larger aperture diameters and increasing capacitances for larger aperture and mesa diameters, resulting in some optimum aperture and mesa diameter.

According to (2.4.70), the *D*-factor is inversely proportional to the square root of the mode volume, and thus, as was also confirmed numerically in Fig. 2.16, to the aperture diameter. This dependence can be clearly obtained from the Fig. 4.28b, where the inverse value of the *D*-factor is shown as a function of the aperture diameter. Again, edge effects for the two VCSELs with the smallest aperture diameters of 4 and 5 µm cause some difference in the values, compared to the linear fit, but the overall trend is unmistakably, additionally confirming high growth and fabrication process quality and predictability of the physical properties for VCSELs with different aperture diameters. The *D*-factor reaches values larger than 9.3 GHz/Sqrt(mA) for the VCSEL with 4 μm aperture diameter and decrease with increasing aperture down to ~2.7 GHz/Sqrt(mA) at 13 μm aperture diameter. Optimization of the active region leaded to much larger *D*-factors for the 980 nm QW-VCSELs as compared to those of the 980 nm SML-VCSELs for nominally identical aperture diameters. For example, the *D*-factor of the 980 nm QW-VCSEL with 5 μm aperture diameter is with ~6.9 GHz/Sqrt(mA) more than twice larger than the *D*-factor of the 5 μm aperture 980 nm SML-VCSEL with ~3.3 GHz/Sqrt(mA). Larger *D*-factors help decisively to reach large bandwidth already at low driving currents, enabling high speed operation in a broader current range, which is an important precondition for good eye quality at higher bit rates.

As can be seen from Fig. 4.29, the maximum bandwidth is hardly dependent on the aperture diameter. The largest bandwidth was measured for the VCSEL with 8 μm aperture diameter and was as large as ~16.2 GHz, which is an improvement by more than 3 GHz as compared to the 980 nm SML-VCSELs, where maximum bandwidths of ~13 GHz were measured. The extracted *D*-factor for this VCSEL was ~4.2 GHz/Sqrt(mA), the extracted *K*-factor was ~0.39 ns. By optimizing device design and epitaxial structure, drastic reduction of electrical parasitic capacitances has been achieved, resulting in the contact pad capacitance C_p of only ~3 fF and the capacitance of the oxide aperture and active region C_a of only ~190 fF for the 8 μm aperture 980 nm QW-VCSEL. Directly comparing the contact pad capacitance C_p of the 980 nm SML-VCSEL with 5 μm aperture diameter, which is ~65 fF, to the contact pad capacitance of the 980 nm QW-VCSEL with nominally the same aperture diameter of 5 μm, which is ~1.3 fF, significant reduction can be obtained, caused by utilizing of the thicker BCB planarization layer and optimizing of the laser structure. The more important capacitance of the oxide aperture and active region C_a was reduced by a factor of two by applying two oxide apertures in the 980 nm QW-VCSELs instead of one single aperture in the 980 nm SML-VCSELs, as can be obtained again by comparison of the values of this capacitance for the 980 nm QW-VCSEL and 980 nm SML-VCSEL with the nominally identical aperture diameter of 5 μm. This capacitance is ~290 fF for the 5 μm aperture 980 nm SML-VCSEL and ~143 fF for the 5 μm aperture 980 nm QW-VCSEL, fully corresponding to the doubling of the thickness of the oxide aperture region.

To investigate high speed properties of the 980 nm QW-VCSELs at elevated temperatures, temperature dependent small signal modulation measurements were carried out for two VCSELs with aperture diameters of 7 and 10 μm. Since the most suitable for high speed and high temperature operation devices are expected to have aperture diameters in between, the temperature dependent behavior of these two VCSELs is important to investigate. Measurements at three characteristic temperatures of 20, 50 and 85°C were carried out and corresponding laser parameters were extracted.

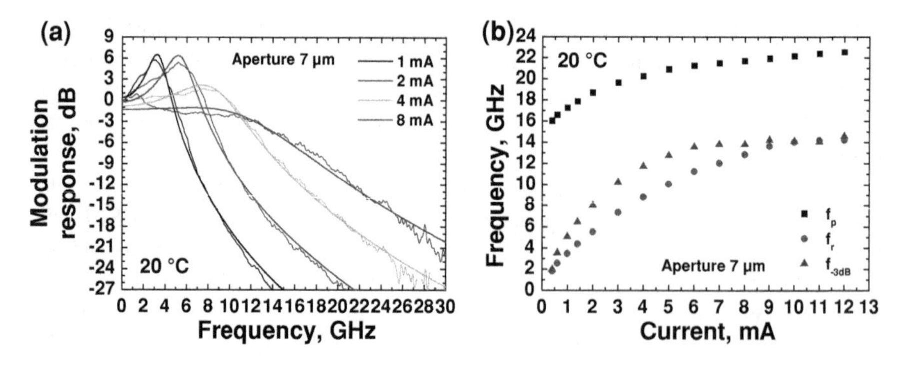

Fig. 4.30 Small signal modulation response with corresponding fits at 20°C for different driving currents (**a**) and extracted values of the parasitic cut-off frequency (*rectangles*), relaxation resonance frequency (*circles*) and 3 dB-frequency (*triangles*) as a function of the current (**b**) for the VCSEL with 7 μm aperture diameter

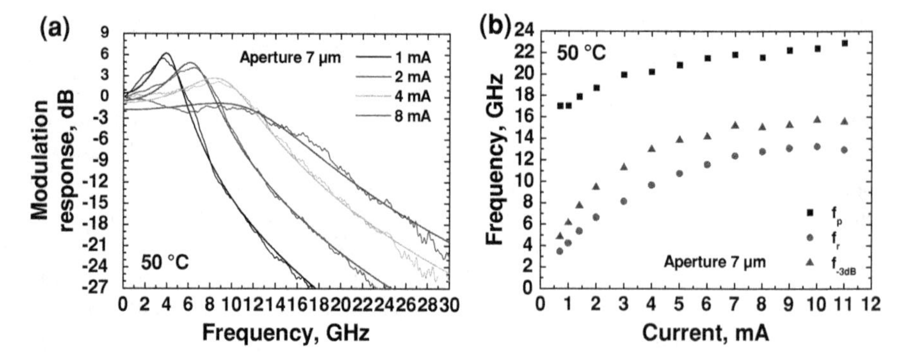

Fig. 4.31 Small signal modulation response with corresponding fits at 50°C for different driving currents (**a**) and extracted values of the parasitic cut-off frequency (*rectangles*), relaxation resonance frequency (*circles*) and 3 dB-frequency (*triangles*) as a function of the current (**b**) for the VCSEL with 7 μm aperture diameter

Figure 4.30 shows measured small signal modulation response at 20°C with corresponding extracted values of the parasitic cut-off frequency, relaxation resonance frequency and 3 dB-frequency.

The room temperature high speed behavior of the 7 μm VCSEL is similar to the 6 μm VCSEL described above. The relaxation resonance frequency reaches values equal to the 3 dB-frequency, showing that the laser is limited mostly by the damping, and thermal effects do not play significant role. Because the parasitic cut-off frequency increases to values larger than 22 GHz, the impact of electrical parasitics is also limited. Maximum bandwidth larger than 14.5 GHz at 20°C was measured.

At higher temperatures the situation changes: relaxation resonance frequency decreases as compared to 20°C and remains at 50 and 85°C below the 3 dB-frequency, as can be obtained from Figs. 4.31 and 4.32. Parasitic cut-off

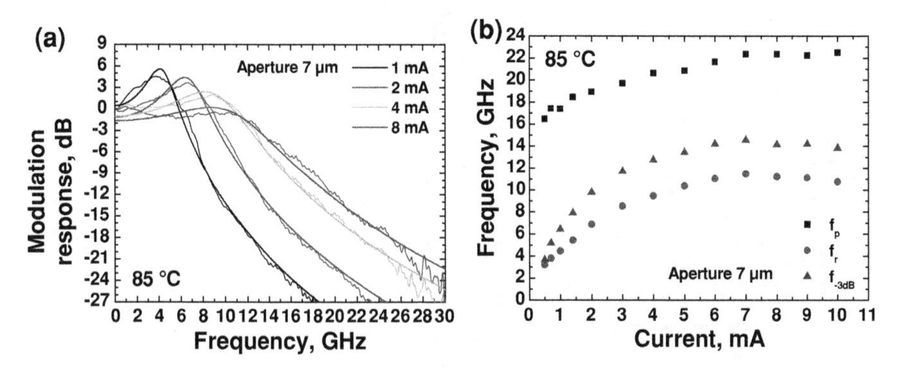

Fig. 4.32 Small signal modulation response with corresponding fits at 85°C for different driving currents (**a**) and extracted values of the parasitic cut-off frequency (*rectangles*), relaxation resonance frequency (*circles*) and 3 dB-frequency (*triangles*) as a function of the current (**b**) for the VCSEL with 7 µm aperture diameter

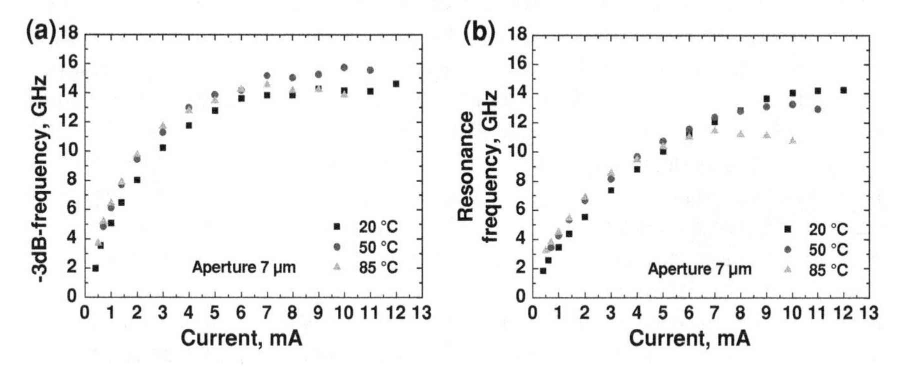

Fig. 4.33 3 dB-frequency (**a**) and relaxation resonance frequency (**b**) at three different temperatures of 20, 50 and 85°C as a function of current for the VCSEL with 7 µm aperture diameter

frequency remains practically unchanged at all three temperatures. At 50°C maximum bandwidth increases to values larger than 15.5 GHz and decreases again at 85°C to values around 14.5 GHz, directly showing the consequence of the gain–cavity detuning, already confirmed by the CW measurements presented previously. The maximum relaxation resonance frequency decreases all the way through from ~14.3 GHz at 20°C to ~13.3 GHz at 50°C and further to ~11.5 GHz at 85°C, corresponding to the temperature increase in the active region at higher ambient temperatures. This decrease does not have a noticeable effect on the bandwidth, because the VCSEL is limited not by the thermal heating but mainly by the damping, as was already mentioned previously.

The excellent temperature stability of the bandwidth can be more clearly recognized by plotting the 3 dB-frequency at all three temperatures in one figure, as shown in Fig. 4.33a. The overall changes in the maximum 3 dB-frequency are

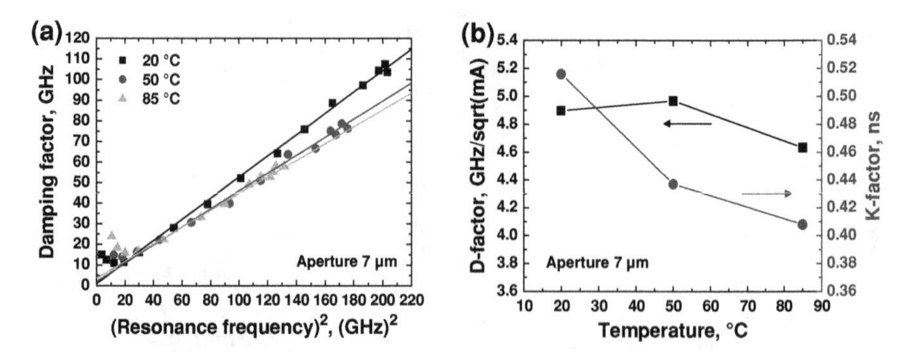

Fig. 4.34 Damping factor as a function of the squared relaxation resonance frequency at 20, 50 and 85°C (**a**) and *D*-factor and *K*-factor as a function of the temperature for the VCSEL with 7 μm aperture diameter

smaller than 1.5 GHz. These values are comparable to those of the 980 nm SML-VCSELs, with the ~2.5 GHz larger maximum bandwidth for the 980 nm QW-VCSELs. In Fig. 4.33b relaxation resonance frequency at different temperature is presented, showing a decrease of the maximum value at higher temperatures.

The increase of the maximum bandwidth at 50°C compared to 20°C is caused additionally by the decrease of the *K*-factor, which can be obtained from the measurement data according to Eq. 2.4.65. In Fig. 4.34a the damping factor as a function of the squared relaxation resonance frequency with corresponding linear fits for different temperatures is shown. The decrease of the *K*-factor with temperature increase can be explained by the increasing absorption loss, which decrease differential efficiency, as was shown in Fig. 4.21b for the same 7 μm aperture VCSEL, and also photon lifetime. The decreased photon lifetime leads to a decrease of the *K*-factor, according to 2.4.67, supporting higher or comparable bandwidths at higher temperatures.

Figure 4.34b demonstrates extracted values of the *D*-factor and the *K*-factor at 20, 50 and 85°C. The *D*-factor increases from ~4.9 GHz/Sqrt(mA) at 20°C to ~5 GHz/Sqrt(mA) at 50°C and then decreases to ~4.6 GHz/Sqrt(mA), while the *K*-factor decreases continuously from ~0.52 ns at 20°C to ~0.44 ns at 50°C and further to ~0.41 ns at 85°C. The combination of the very temperature stable *D*-factor, small effect of electrical parasitics and weak effect of thermal heating leads, together with decreasing damping at higher temperatures, to the very temperature insensitive bandwidth in the whole temperature range from 20 up to 85°C, as shown in Fig. 4.33a.

Temperature dependent small signal modulation measurements with corresponding fits and extracted values of the parasitic cut-off frequency, relaxation resonance frequency and 3 dB-frequency for the 980 nm QW-VCSEL with 10 μm aperture diameter for temperatures of 20, 50 and 85°C are shown in Figs. 4.35, 4.36 and 4.37, correspondingly.

The high speed temperature behavior of the 10 μm aperture device is in some respect different from the 7 μm aperture VCSEL. Already at 20°C the relaxation

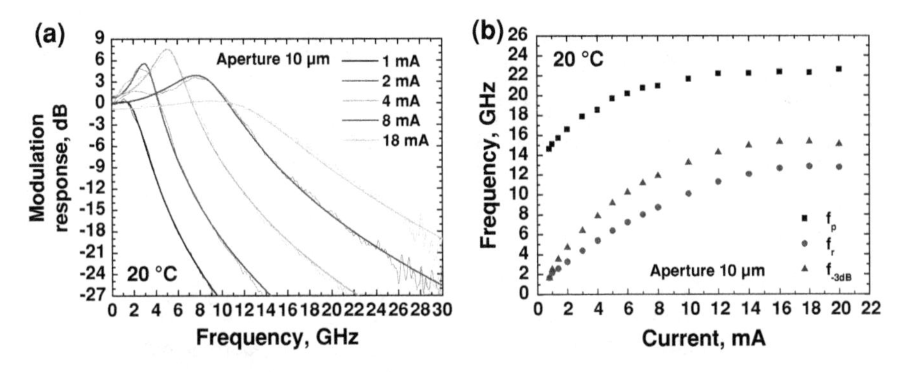

Fig. 4.35 Small signal modulation response with corresponding fits at 20°C for different driving currents (**a**) and extracted values of the parasitic cut-off frequency (*rectangles*), relaxation resonance frequency (*circles*) and 3 dB-frequency (*triangles*) as a function of the current (**b**) for the VCSEL with 10 µm aperture diameter

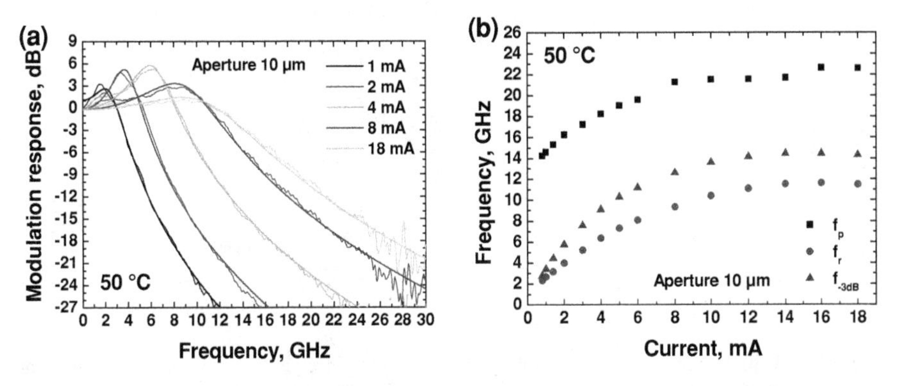

Fig. 4.36 Small signal modulation response with corresponding fits at 50°C for different driving currents (**a**) and extracted values of the parasitic cut-off frequency (*rectangles*), relaxation resonance frequency (*circles*) and 3 dB-frequency (*triangles*) as a function of the current (**b**) for the VCSEL with 10 µm aperture diameter

resonance frequency is at all currents smaller than the 3 dB-frequency, showing a larger effect of the thermal heating. Nevertheless, maximum bandwidth of ~15.4 GHz was achieved at 20°C. The effect of electrical parasitics is also for this device small at all temperatures, which follows from the temperature insensitive maximum parasitic cut-off frequency of larger than 22 GHz at all three temperatures.

The increased role of the thermal effects can be more clearly recognized at higher temperatures, as shown in Figs. 4.36 and 4.37. At 50 and 85°C the difference between the relaxation resonance frequency and the 3 dB-frequency increases. The maximum bandwidth measured at 50°C was ~14.5 GHz and ~13.2 GHz at 85°C.

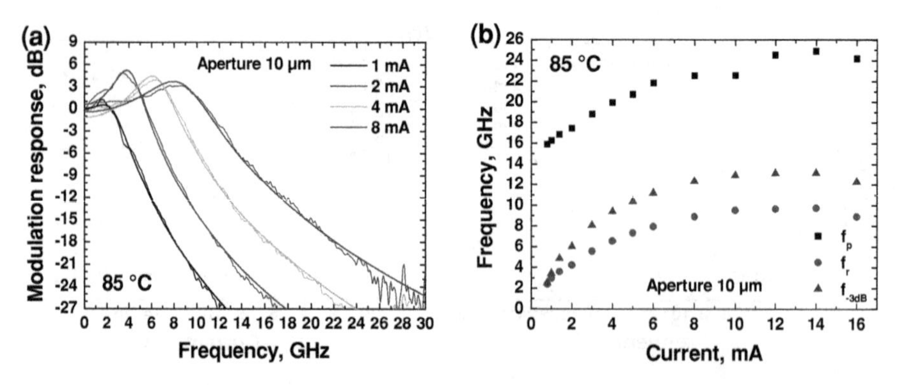

Fig. 4.37 Small signal modulation response with corresponding fits at 85°C for different driving currents (**a**) and extracted values of the parasitic cut-off frequency (*rectangles*), relaxation resonance frequency (*circles*) and 3 dB-frequency (*triangles*) as a function of the current (**b**) for the VCSEL with 10 μm aperture diameter

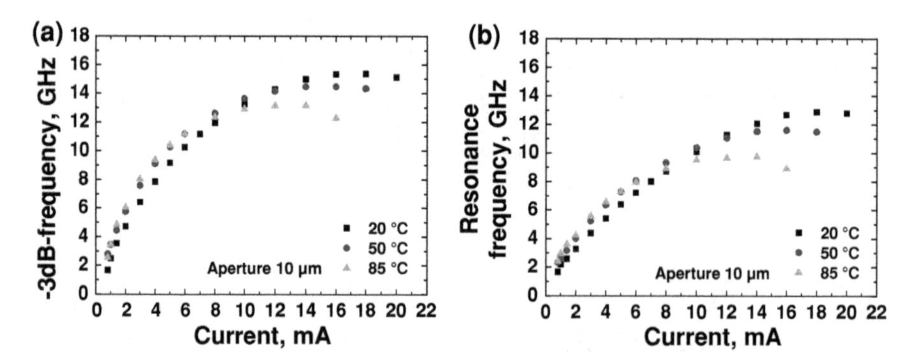

Fig. 4.38 3 dB-frequency (**a**) and relaxation resonance frequency (**b**) at three different temperatures of 20, 50 and 85°C as a function of current for the VCSEL with 10 μm aperture diameter

The decrease of the 3 dB-frequency and of the relaxation resonance frequency with the temperature can be more clearly obtained from the Fig. 4.38, where these parameters for 20, 50 and 85°C are shown. Thus, the temperature stability of the 10 μm aperture VCSEL is weaker as compared to the 7 μm aperture device. Nevertheless, the overall changes of the maximum 3 dB-frequency in the temperature range from 20 up to 85°C are as small as ~2.2 GHz, again showing reasonable good temperature insensitivity of the high speed properties of the 10 μm aperture VCSEL.

The temperature behavior of the *D*-factor and the *K*-factor of the 10 μm aperture VCSEL is similar to those of the 7 μm aperture device and is shown in Fig. 4.39b. The *D*-factor increases from ~3.48 GHz/Sqrt(mA) at 20°C to ~3.53 GHz/Sqrt(mA) at 50°C and then decreases to ~3.35 GHz/Sqrt(mA) at 85°C, directly corresponding to the minimum of the threshold current around 50°C

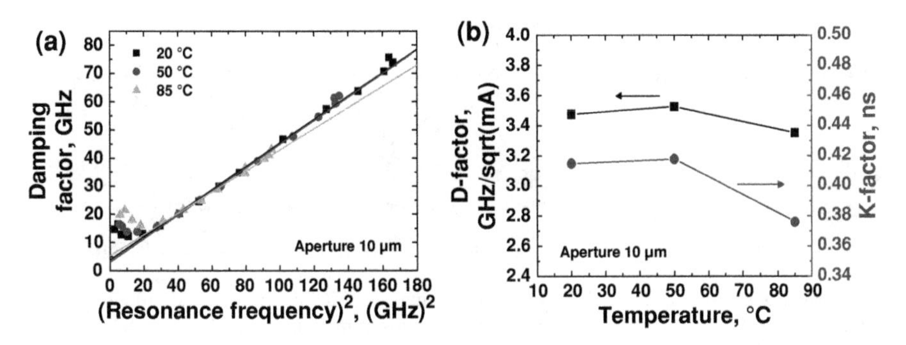

Fig. 4.39 Damping factor as a function of the squared relaxation resonance frequency at 20, 50 and 85°C (**a**) and *D*-factor and *K*-factor as a function of the temperature for the VCSEL with 10 μm aperture diameter

shown in Fig. 4.22b and caused by the gain–cavity detuning. The *K*-factor remains practically constant at ~0.42 ns from 20 up to 50°C and then decreases to ~0.38 ns at 85°C, as can be already seen in Fig. 4.39a.

Since all investigated physical properties have shown a continuously change with the aperture diameter, one can conclude that devices with aperture diameters between 7 and 10 μm have very temperature insensitive high speed characteristics, e.g., bandwidth, relaxation resonance frequency, *D*-factor, *K*-factor and parasitic cut-off frequency. Combined together with the very temperature stable CW characteristics of these devices presented in the previous section, it demonstrates a great potential and suitability of the 980 nm QW-VCSELs for the high speed high temperature stable data transmission and successfulness of the applied optimization concepts, both for the epitaxial structure and for the laser design. In the next section large signal modulation experiments carried out on 980 nm QW-VCSELs at different temperatures will be presented.

4.2.4 Large Signal Modulation Characteristics

Large signal modulation experiments using a non-return to zero (NRZ) data format with a (2^7-1) pseudorandom binary sequence (PRBS) in a standard back-to-back measurement configuration (BTB, ~3 m fiber) were performed with an SHF 12100 B bit pattern generator and a New Focus 1434-50-M photodetector on 980 nm QW-VCSELs with different aperture diameters, in order to proof the ability of the investigated lasers to transfer data at high speed at different temperatures. Best results were achieved for devices with aperture diameters between 7 and 10 μm.

In Fig. 4.40 the measured optical eye diagrams for data bit rates of 20, 25 and 30 Gbit/s at temperature of 25°C along with the corresponding forward drive currents and extracted signal-to-noise ratios (S/N) are shown for the VCSEL with 7 μm aperture diameter. The peak-to-peak drive voltage $V_{p\text{–}p}$ was for all eye

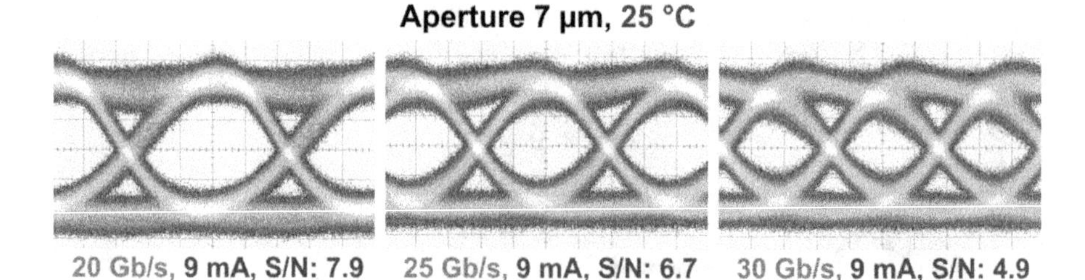

Fig. 4.40 Back-to-back, 2^7-1 PRBS eye diagrams at 25°C at 20, 25 and 30 Gbit/s with corresponding driving currents and signal-to-noise ratios at modulation voltage of 0.88 V for the VCSEL with 7 µm aperture diameter

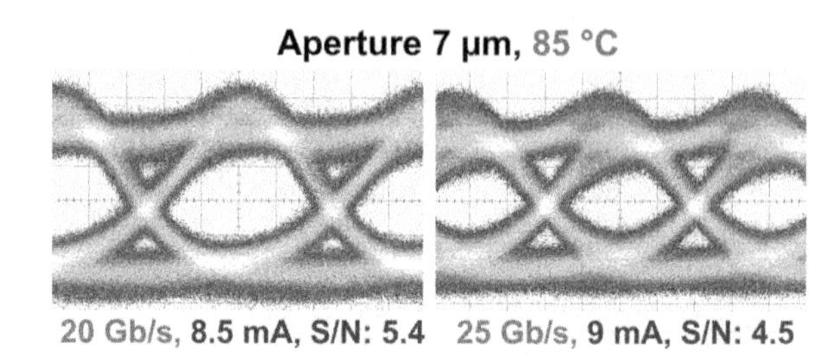

Fig. 4.41 Back-to-back, 2^7-1 PRBS eye diagrams at 85°C at 20 and 25 Gbit/s with corresponding driving currents and signal-to-noise ratios at modulation voltage of 0.88 V for the VCSEL with 7 µm aperture diameter

diagrams 0.88 V. All eyes are clearly open. The signal-to-noise ratio is ~7.9 at 20 Gbit/s, ~6.7 at 25 Gbit/s and ~4.9 at 30 Gbit/s. Extinction ratios of all measured eye diagrams were around 4 dB. The measured clearly open eye diagrams at bit rates up to 30 Gbit/s agree very good to the measured small signal modulation bandwidths in the range of 14–15 GHz for the 7 µm aperture VCSEL presented in the previous section.

At the temperature of 85°C the VCSEL demonstrates clearly open eye diagrams at bit rates up to 25 Gbit/s, as shown in Fig. 4.41. Signal-to-noise ratio of ~5.4 was measured at 85°C for the bit rate of 20 Gbit/s and for 25 Gbit/s it was ~4.5. Although at higher temperatures the bandwidth remains practically unchanged, because of the output power decrease the signal-to-noise ratio decreases, preventing VCSEL operation at 30 Gbit/s at 85°C. The eyes at 85°C have a small rising-edge overshoot, which is caused by the reduced damping, as discussed in the previous section. The clearly open eye diagram at 25°C at bit rates up to 30 Gbit/s and at 85°C at bit rates up to 25 Gbit/s unmistakably demonstrates the suitability of the VCSEL to transfer data with such high frequencies at room temperature and also at 85°C.

The best optical eye diagrams at different temperatures were obtained for the VCSEL with 8 µm aperture diameter. Figure 4.42 demonstrates the measured eye diagrams at 25°C for different bit rates of 20, 25 and 30 Gbit/s for the 8 µm aperture VCSEL.

All eyes are clearly open up to bit rates of 30 Gbit/s. Signal-to-noise ratios of ~8.9 at 20 Gbit/s, ~7.4 at 25 Gbit/s and ~5.3 at 30 Gbit/s were measured,

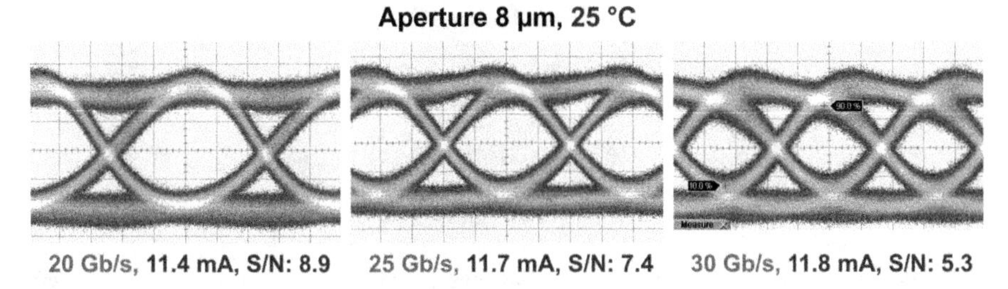

Fig. 4.42 Back-to-back, 2^7-1 PRBS eye diagrams at 25°C at 20, 25 and 30 Gbit/s with corresponding driving currents and signal-to-noise ratios at modulation voltage of 0.88 V for the VCSEL with 8 μm aperture diameter

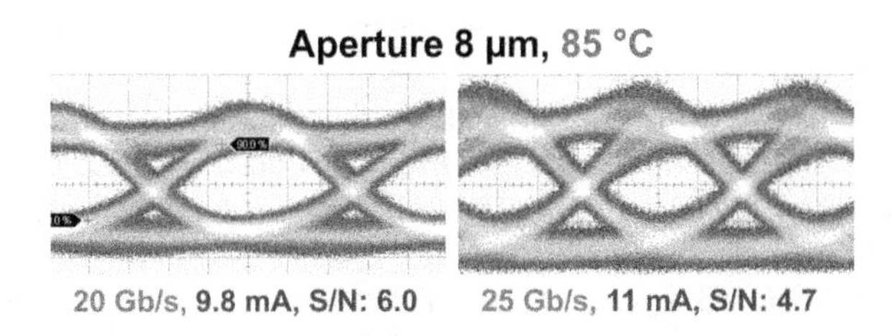

Fig. 4.43 Back-to-back, 2^7-1 PRBS eye diagrams at 85°C at 20 and 25 Gbit/s with corresponding driving currents and signal-to-noise ratios at modulation voltage of 0.75 V for the VCSEL with 8 μm aperture diameter

which are larger as compared to the values measured with the 7 μm aperture VCSEL. As in the case of the 7 μm VCSEL, the eye diagrams of the 8 μm VCSEL at 20 and 25 Gbit/s clearly demonstrate plateaus corresponding to the "0" and "1" signal levels. This fact unmistakably confirms that the quality of the eye diagrams at data rates of 20 and 25 Gbit/s is not limited by the laser speed. These plateaus can be also obtained at these data rates at the temperature of 85°C for both devices, as shown in Figs. 4.41 and 4.43, confirming again that the eye quality at elevated temperatures at data rates up to 25 Gbit/s is limited not by the laser speed but rather by the decreased optical output power and thus reduced signal-to-noise ratio. At 30 Gbit/s the plateaus vanish and the eye diagram quality becomes affected also by the limited speed of the VCSEL, which practically perfectly corresponds to the maximum small signal modulation bandwidths of ~14–16 GHz presented in the previous section.

Measured eye diagrams at 85°C at bit rates of 20 and 25 Gbit/s for the 8 μm VCSEL together with the corresponding driving currents and signal-to-noise ratios are shown in Fig. 4.43. The peak-to-peak modulation voltage V_{p-p} was for these eye diagrams 0.75 V, which is smaller as compared to the voltage for the eye diagrams measured at 25°C. The signal-to-noise ratios are also here larger as compared to the 7 μm VCSEL, and are ~6 at 20 Gbit/s and ~4.7 at 25 Gbit/s. The S/N of the 20 Gbit/s 85°C eye is comparable to the S/N of the 980 nm SML-VCSELs at the same bit rate of 20 Gbit/s and the same temperature of 85°C.

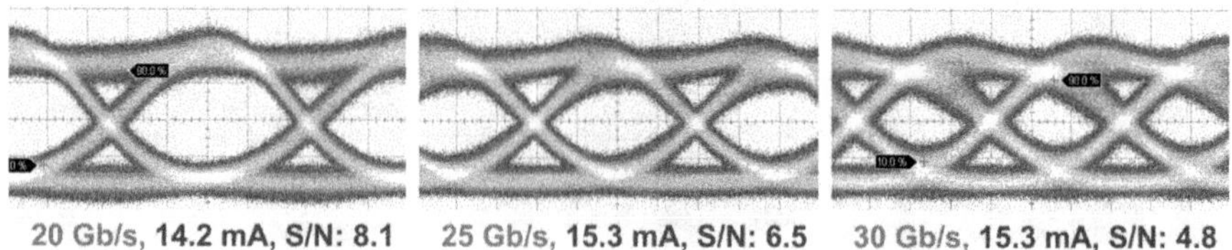

Fig. 4.44 Back-to-back, 2^7-1 PRBS eye diagrams at 25°C at 20, 25 and 30 Gbit/s with corresponding driving currents and signal-to-noise ratios at modulation voltage of 0.75 V for the VCSEL with 10 μm aperture diameter

But because of the increased bandwidth the 980 nm QW-VCSELs are able to operate at 25 Gbit/s at elevated temperatures up to 85°C, which was not the case for the 980 nm SML-VCSEL.

Finally, Fig. 4.44 demonstrates optical eye diagrams measured at 25°C at data rates of 20, 25 and 30 Gbit/s with the 10 μm aperture VCSEL. Corresponding driving currents and S/Ns are shown as well. The modulation voltage $V_{p\text{-}p}$ was 0.75 V for all three eye diagrams.

The room temperature eye diagrams of the 10 μm aperture VCSEL are very similar to the eye diagrams measured at 25°C with VCSELs with 7 and 8 μm aperture diameters. Signal-to-noise ratios of ~8.1 at 20 Gbit/s, ~6.5 at 25 Gbit/s and ~4.8 at 30 Gbit/s were registered. But because of the reduced bandwidth at 85°C as compared to the 7 and apparent to the 8 μm aperture devices (Figs. 4.33 and 4.38), the 10 μm aperture VCSEL was limited to bit rates of 20 Gbit/s at 85°C.

The measured optical eye diagrams of the investigated lasers practically perfectly agree to the measured small signal modulation bandwidths and fully confirm the suitability of the 980 nm QW-VCSELs to operate at bit rates up to 30 Gbit/s at 25°C and up to 25 Gbit/s at 85°C.

4.2.5 Summary of the 980 nm QW-VCSEL Results

CW, small and large signal modulation experiments have demonstrated an excellent temperature stability of the 980 nm QW-VCSELs from room temperature up to 85°C. Very temperature stable CW characteristics and very temperature insensitive small and large signal modulation characteristics were measured for VCSELs with different oxide aperture diameters, showing consistent values of the physical parameters as a function of the aperture size. This fact confirmed the very high quality of the growth and fabrication processes. Small signal modulation bandwidths larger than 16 GHz at 25°C were obtained. The overall bandwidth change with temperature in the range from 25 up to 85°C was limited to only ~1.5 GHz for the best devices. Consequently, very temperature stable and clearly open optical eye diagrams at bit rates up to 25 Gbit/s and temperatures from 25 up to 85°C were measured. At 25°C the eye diagrams remain open even at bit rates up to 30 Gbit/s.

The excellent temperature stability of the 980 nm QW-VCSELs was achieved by applying several additional optimization concepts, which were developed based on the results of the 980 nm SML-VCSELs presented previously. First of all, drastic reduction of both parasitic capacitances C_p and C_a was achieved by applying an even thicker BCB layer, double oxide apertures and choosing smaller mesa diameters for the same aperture sizes. The contact pad parasitic capacitance was reduced from several tens of fF to only several fF, making it practically negligible. The capacitance of the active region and oxide apertures was decreased by more than a factor of two for nominally the same aperture diameter, and was as low as ~143 fF for the 5 μm aperture 980 nm QW-VCSEL. The decreased parasitic capacitances have lead to improvement of the parasitic cut-off frequency to values larger than 22 GHz for devices with different aperture diameter.

The second major improvement was achieved by optimization of the active region. Applying five highly strained 4.2 nm thick $In_{0.21}Ga_{0.79}As$ QWs separated and surrounded by the 6 nm thick $GaAs_{0.88}P_{0.12}$ strain compensation layers resulted in an increase of the differential gain, which can be directly obtained by comparison of the *D*-factor for nominally the same oxide aperture diameters. While in the case of the 980 nm SML-VCSELs the *D*-factor of the 5 μm aperture VCSEL at room temperature was only ~3.3 GHz/Sqrt(mA), in 5 μm aperture 980 nm QW-VCSEL the *D*-factor was as high as ~6.9 GHz/Sqrt(mA), which is by more than a factor of two larger, leading to a much faster increase of the bandwidths with driving current and reducing the effect of thermal heating. This has enabled to reach high bandwidths at larger aperture diameters, reducing the operation current density and increasing device reliability. Also the utilizing of the gain peak–cavity resonance detuning of nominally 15 nm has decisively contributed to the excellent temperature stability of the 980 nm QW-VCSELs and was confirmed by the CW and small signal modulation measurements, showing a minimum of the threshold current and a maximum of the *D*-factor at temperatures around 50°C.

From the small signal modulation experiments also limiting factors of the high speed laser operation were obtained. The major limit is represented by the damping. The corresponding *K*-factors were comparable for both 980 nm QW-VCSELs and 980 nm SML-VCSELs. To reduce damping the photon lifetime should be reduced and the differential gain increased further, accordingly to (2.4.67). The first can be achieved by reducing the number of the top DBR pairs, which will simultaneously increase the differential efficiency, accordingly to (2.4.19) and (2.4.20), and thus the output power. Since the limited output power was a further limitation of the 980 nm QW-VCSELs, especially at high temperatures, reduction of the reflectivity of the top mirror will help also to overcome it. Logically, an optimum value of the top mirror reflectivity should be aimed, in order to keep the overall losses reasonable. To increase the differential gain, thicker QWs could be applied, leading to an increased of the optical confinement factor Γ and thus decreased threshold gain g_{th} and threshold current density N_{th}, accordingly to (2.4.23) and (2.4.25). The decreased threshold current density will result in the increase of the differential gain a, according to (2.4.36).

All of the described optimization concepts were applied in the 850 nm QW-VCSELs, which will be presented in the next chapter, and have led to the error free laser operation at the record high data rate of 38 Gbit/s.

References

1. Doany FE, Schares L, Schow CL, Schuster C, Kuchta DM, Pepeljugoski PK (2006) Chip-to-chip optical interconnects. OFC, Anaheim, CA, USA, OFA3
2. Kam DG, Ritter MB, Beukema TJ, Bulzacchelli JF, Pepeljugoski PK, Kwark YH, Shan L, Gu X, Baks CW, John RA, Hougham G, Schuster C, Rimolo-Donadio R, Wu B (2009) Is 25 Gb/s on-board signaling viable? IEEE Trans Adv Packag 32(2):328–344
3. Kash JA, Doany F, Kuchta D, Pepeljugoski P, Schares L, Schaub J, Schow C, Trewhella J, Baks C, Kwark Y, Schuster C, Shan L, Parel C, Tsang C, Rosner J, Libsch F, Budd R, Chiniwall P, Guckenberger D, Kucharski D, Dangel R, Offrein B, Tan M, Trott G, Lin D, Tandon A, Nystrom M (2006) Terabus: a chip-to-chip parallel optical interconnect. OFC, Anaheim, CA, USA, TuW3
4. Bimberg D (2008) Quantum dot based nanophotonics and nanoelectronics. IEEE Electron Lett 44(33):168–171
5. Krestnikov IL, Ledentsov NN, Hoffmann A, Bimberg D (2008) Arrays of two-dimensional islands formed by submonolayer insertions: growth, properties, devices. Phys Stat Sol (a) 183(2):207–233
6. Bressler-Hill V, Lorke A, Varma S, Petroff PM, Pond K, Weinberg WH (1994) Initial stages of InAs epitaxy on vicinal GaAs(001)-(2 × 4). Phys Rev B 50:8479–8487
7. Zhukov E, Kovsh AR, Mikhrin SS, Maleev NA, Ustinov VM, Livshits DA, Tarasov IS, Bedarev DA, Maximov MV, Tsatsul'nikov AF, Soshnikov IP, Kop'ev PS, Alferov ZhI, Ledentsov NN, Bimberg D (1999) 3.9 W CW power from sub-monolayer quantum dot diode laser. Electron Lett 35(21):1845–1847
8. Mikhrin SS, Zhukov AE, Kovsh AR, Maleev NA, Ustinov VM, Shernyakov Yu M, Soshnikov IP, Livshits DA, Tarasov IS, Bedarev DA, Volovik BV, Maximov MV, Tsatsul'nikov AF, Ledentsov NN, Kop'ev PS, Bimberg D, Alferov Zh I (2000) 0.94 μm diode lasers based on Stranski–Krastanow and sub-monolayer quantum dots. Semicond Sci Technol 15:1061–1064
9. Pötschke K (2009) Untersuchungen zur Bildung von Quantenpunkten im Stranski-Krastanov und im submonolagen Wachstumsmodus. Thesis, TU Berlin
10. Mikhaelashvili V, Tessler N, Nagar R, Eisenstein G, Dentai AG, Chandrasakhar S, Joyner CH (1994) Temperature dependent loss and overflow effects in quantum well lasers. IEEE Photon Technol Lett 6(11):1293–1296
11. Chang Y-C, Coldren LA (2009) Efficient, high-data-rate, tapered oxide-aperture vertical-cavity surface-emitting lasers. IEEE J Sel Top Quantum Electron 15(3):704–715
12. Safaisini R, Joseph JR, Louderback D, Jin X, Al-Omari AN, Lear KL (2008) Temperature dependence of 980-nm oxide-confined VCSEL dynamics. IEEE Photon Technol Lett 20(14):15
13. Coldren LA, Corzine SW (1995) Diode lasers and photonic integrated circuits. Wiley, New York

Chapter 5
High Speed 850 nm VCSEL Results

For future applications of semiconductor lasers in LAN/SAN, to increase the maximum achievable bit rate at room temperature becomes the ultimate goal and challenge, while the importance of the high temperature stability of the laser properties plays a smaller role, as compared to application for on-chip, chip-to-chip and module-to-module data transmission. From the other side, there are optical standards which define the wavelength of 850 nm to be used for data transmission in LAN/SAN, thus the freedom to choose any desired wavelength is no more present. Consequently, to meet the requirements for the future LAN/SAN optical data transmission, cheap and robust VCSELs emitting around 850 nm with the highest possible data transmission bit rate should be realized. According to the coming standards, lasers with the bit rate of 30 Gbit/s and larger are necessary in the next several years. Additionally, optics is coming also to shorter distances and in the near future definitively will replace copper-based data transmission lines for such application like, for example, data communication between a personal computer and portable devices (USB stick, mp3 player, etc.), between computer and monitor, between TV and player, etc. These application fields require semiconductor lasers capable to a very cheap mass production, and VCSELs are practically the only suitable candidates.

Following this motivation we have developed 850 nm QW-VCSELs optimized to achieve the largest possible bit rate at room temperature. Additionally to the optimizations already applied for the 980 nm QW-VCSELs described in the previous chapter, thicker QWs have been utilized to increase the optical confinement factor and the number of the top DBR pairs was reduced to decrease the damping and to increase the differential efficiency and the output power. These optimization steps have lead to worldwide first 850 nm VCSELs operating error-free at the bit rate of 38 Gbit/s at room temperature. These were also first of any oxide-confined VCSELs operating at such high bit rates. At the elevated temperature of 85°C error-free transmission at 25 Gbit/s has been measured, repeating the worldwide record value measured few months earlier [1].

A. Mutig, *High Speed VCSELs for Optical Interconnects*, Springer Theses,
DOI: 10.1007/978-3-642-16570-2_5,

5.1 Device Structure

Since the MOCVD growth technique is advantageously for cheap mass production of photonic devices, the growth of the 850 nm QW-VCSELs was carried out by MOCVD. Semi-insulating GaAs (100) substrates were used to reduce electrical parasitics. Five compressively strained $In_{0.05}Ga_{0.95}As$ QWs were applied as active region to increase differential gain. The thickness of the QWs was with 4.8 nm thicker in the 850 nm QW-VCSELs as 4.2 nm used in the 980 nm QW-VCSELs, which has increased the optical confinement factor. The thickness of the $Al_{0.30}Ga_{0.70}As$ barriers utilized in the 850 nm QW-VCSELs was with 5 nm thinner as the thickness of the $GaAs_{0.88}P_{0.12}$ barriers used in the 980 nm QW-VCSELs, which were 6 nm thick. This thickness reduction has also contributed to the increase of the optical confinement factor. With the described improvements, the calculated one-dimensional optical confinement factor in the 850 nm QW-VCSELs was larger than 3.8%, which is more than 25% larger than the calculated optical confinement factor in the 980 nm QW-VCSELs, which was 3.0%. This improvement is decisive for the high speed laser properties and can be also used to partially compensate increased losses because of the reduced number of the top DBR mirror pairs, without sacrificing low threshold carrier density. Since the 850 nm QW-VCSELs were optimized for room temperature operation, gain peak–cavity resonance detuning was reduced to nominally 8 nm, in contrast to 15 nm applied in the high temperature stable 980 nm QW-VCSELs. The reduced gain peak–cavity resonance detuning helps additionally to keep the threshold carrier density reasonable low for a large differential gain, accordingly to (2.4.36). The QWs with barriers were surrounded by the two 20 nm thick $Al_{0.30}Ga_{0.70}As$ barrier layers to improve carrier localization in the active region.

Similar to the 980 nm QW-VCSELs described in the previous chapter, two 30 nm thick $Al_{0.98}Ga_{0.02}As$ aperture layers lying in the field intensity nodes have been introduced for optical and electrical guiding and for reduction of the electrical parasitic capacitances. The first aperture was in the first top DBR pair closest to the cavity region, the second aperture was inside the cavity just below the first DBR pair, both positioned similar to the 980 nm QW-VCSELs. The cavity length was 3/2 λ.

Optimized doped $Al_{0.90}Ga_{0.10}As/Al_{15}Ga_{0.85}As$ DBR mirrors were applied for both top and bottom mirror. The top p-doped DBR contained 22.5 mirror pairs, while the bottom n-doped DBR consisted of 35 mirror pairs. Linear gradings with 20 nm thickness were applied to reduce electrical resistance of the mirrors, similar to the 980 nm QW-VCSELs.

High speed 850 nm QW-VCSELs with co-planar top contacts and different aperture diameters were fabricated, using the same fabrication process scheme as for the 980 nm SML-VCSELs and 980 nm QW-VCSELs described in the previous chapter. To decrease parasitic capacitances even further, oxidation length has been further reduces, so that VCSELs with nominally 9 μm aperture diameter were realized not at 32 μm mesa diameter, as was the case in the already optimized

980 nm QW-VCSELs, but at smaller mesa diameter of 28 μm. Fields with VCSELs with different aperture diameters were fabricated, enabling the search for the optimum aperture size. Devices with aperture diameters between 6 and 9 μm have shown the best high speed performance at relatively low current densities, and their CW and high speed measurement results are presented in the next sections of this chapter.

5.2 Static Characteristics

Since the most suitable for high speed operation 850 nm QW-VCSELs were devices with aperture diameters between 6 and 9 μm, in this section CW characteristics of three devices with aperture diameter of 6, 7 and 9 μm will be presented. Static characteristics of the 850 nm QW-VCSEL with 6 μm aperture diameter at different temperatures between 20 and 100°C are shown in Fig. 5.1 together with extracted values of the maximum differential efficiency and threshold current.

The maximum output power is ~5.3 mW at 20°C and decreases to ~3.2 mW at 80°C and further to ~2.5 mW at 100°C. The change of the output power with the temperature is comparable to both 980 nm SML-VCSELs and 980 nm QW-VCSELs. In all three cases the value of the maximum power at 80 or 85°C was about ~60% of the value at 20 or 25°C. The maximum differential efficiency of the 6 μm 850 nm QW-VCSEL is ~0.89 W/A at 20°C and decreases to ~0.74 W/A at 80°C, and further to ~0.67 W/A at 100°C. These values are larger than the values of the maximum differential efficiency of the 980 nm QW-VCSELs, because of the reduced number of the top mirror pairs and also decreased internal losses, arising among other from the shorter wavelength of 850 nm as compared to 980 nm. Again, the value of the maximum differential efficiency at 80°C is about 75% of the value at 20°C, which is pretty comparable to the changes in the 980 nm SML-VCSELs and 980 nm QW-VCSELs. Comparable changes of the differential

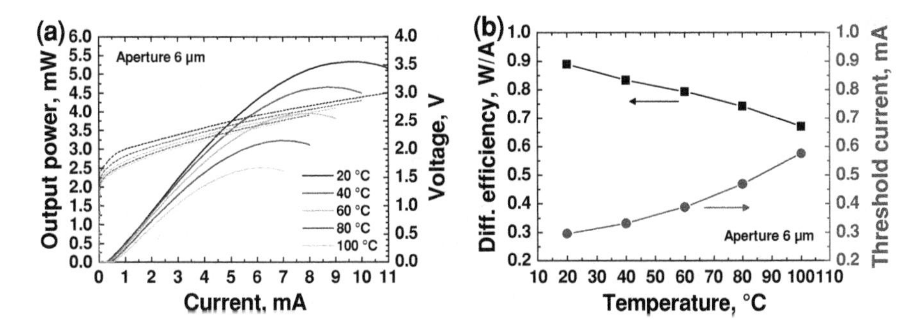

Fig. 5.1 L–U–I characteristics (**a**) and extracted values of the maximum differential efficiency and threshold current (**b**) at different temperatures between 20 and 100°C for the VCSEL with 6 μm aperture

efficiency with the temperature for all investigated VCSELs indicate comparable changes of the internal losses with temperature increase, according to (2.3.15). This fact contributes also to the similar changes of the maximum output power of all investigated VCSELs. The large difference of the 850 nm QW-VCSELs, as compared to both types of the 980 nm lasers, can be obtained by investigation of the temperature behavior of the threshold current, which is shown for the 6 μm aperture 850 nm QW-VCSEL in Fig. 5.1b as well. The threshold current of the 6 μm laser is ~295 μA at 20°C and increases to ~470 μA at 80°C and further to ~575 μA at 100°C. Because of the smaller value of the gain–cavity detuning of nominally 8 nm no minimum in the threshold current as a function of temperature can be obtained. Also the increase of the threshold current between 20 and 80°C is with 60% of the value at 20°C much larger than corresponding changes of the threshold current for the 980 nm VCSELs. This difference is caused by the larger gain–cavity detuning of nominally 15 nm for both types of the 980 nm VCSELs described in the previous chapter. Logically, the stability of the high speed properties of the 850 nm QW-VCSELs is expected to be lower as compared to 980 nm devices, as will be confirmed by the measurements presented in the following sections. From the other side, because of the optimized active region and device design, the room temperature bandwidth of the 850 nm QW-VCSELs is expected to be larger as those of the 980 nm lasers, which also will be confirmed by the high speed modulation measurements. The differential resistance of the 6 μm 850 nm QW-VCSEL was ~110 Ω at 5 mA and 20°C. The thermal resistance of the device was ~4.4 K/mW, comparable to the values of the corresponding 980 nm QW-VCSELs.

In Fig. 5.2 the L–U–I characteristics and extracted values of the maximum differential efficiency and threshold current at different temperatures for the 850 nm QW-VCSEL with aperture diameter of 7 μm are shown.

The maximum output power is ~5.7 mW at 20°C, decreases to ~3.3 mW at 80°C and is ~2.5 mW at 100°C. The maximum differential efficiency is ~0.74 W/A at 20°C and decreases to ~0.63 W/A at 80°C, and further to ~0.58 W/A at 100°C. The threshold current is ~365 μA at 20°C and increases to ~600 μA at 80°C, and reaches ~740 μA at 100°C. The temperature behavior of the static

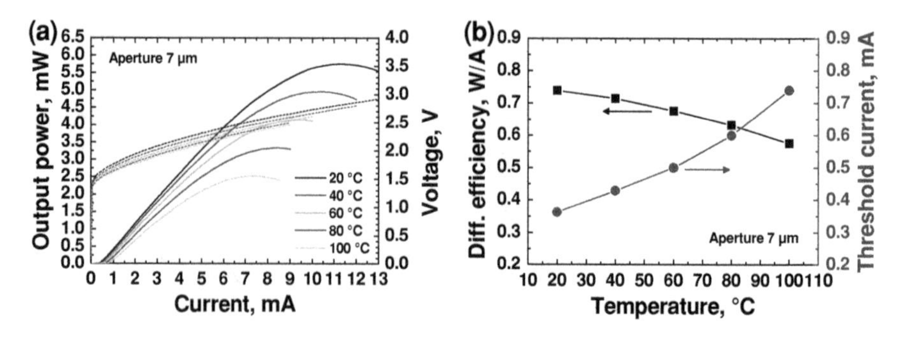

Fig. 5.2 L–U–I characteristics (**a**) and extracted values of the maximum differential efficiency and threshold current (**b**) at different temperatures between 20 and 100°C for the VCSEL with 7 μm aperture

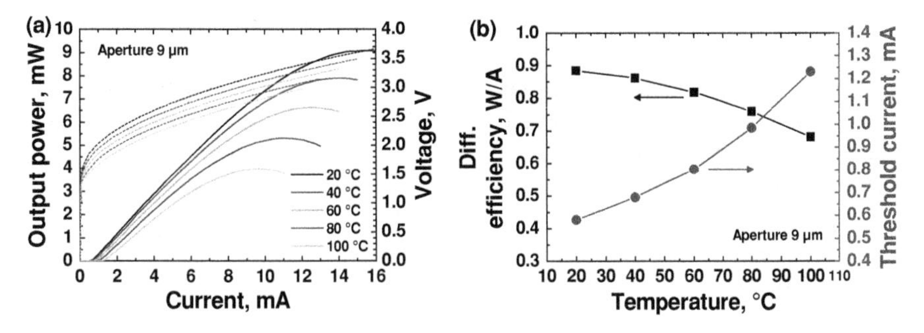

Fig. 5.3 L–U–I characteristics (**a**) and extracted values of the maximum differential efficiency and threshold current (**b**) at different temperatures between 20 and 100°C for the VCSEL with 9 μm aperture

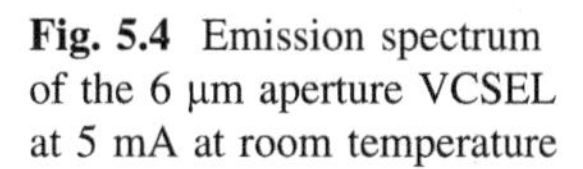
Fig. 5.4 Emission spectrum of the 6 μm aperture VCSEL at 5 mA at room temperature

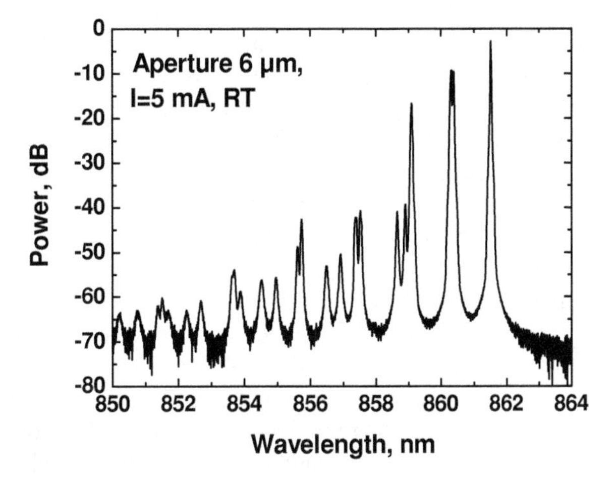

characteristics of the 7 μm aperture VCSEL is similar to the behavior of the 6 μm aperture device. The differential resistance of the 7 μm aperture device was ~95 Ω at 6 mA at 20°C.

Static characteristics of the 850 nm QW-VCSEL with 9 μm aperture diameter with corresponding values of the maximum differential efficiency and threshold current are shown in Fig. 5.3 for different temperatures between 20 and 100°C.

The maximum output power reaches values larger than 9 mW at 20°C and decreases to ~5.3 mW at 80°C, and is still ~4 mW at 100°C. The maximum differential efficiency is ~0.89 W/A at 20°C, decreases to ~0.76 W/A at 80°C, and is ~0.68 W/A at 100°C. The threshold current increases from ~580 μA at 20°C to ~985 μA at 80°C, and reaches ~1.23 mA at 100°C. The differential efficiency of the 9 μm VCSEL was ~90 Ω at 9 mA at 20°C. The thermal resistance of the 9 μm aperture VCSEL was ~2.2 K/mW.

All investigated 850 nm VCSELs had aperture diameters larger than 5 μm to achieve low operation current density for better device reliability, and consequently were multimode. In Fig. 5.4 the emission spectrum of the 6 μm aperture laser at 5 mA and room temperature is shown, clearly demonstrating a multimode operation.

Measured CW characteristics have demonstrated large output powers, high differential efficiencies and low enough threshold currents of the investigated 850 nm QW-VCSELs, which build a good basis for proper high speed laser performance.

5.3 Small Signal Modulation Analysis

To characterize physical processes important for high speed operation, small signal modulation measurements were performed at different temperatures at frequencies up to 40 GHz and the fundamental physical parameters have been extracted. The analysis was carried out for two devices with oxide aperture diameter of nominally 6 and 9 μm to investigate different types of high speed laser behavior. In Fig. 5.5a measured small signal modulation response curves at 25°C at different driving currents together with the corresponding fits are presented for the 850 nm QW-VCSEL with 6 μm aperture diameter. Fig. 5.5b shows corresponding extracted values of the parasitic cut-off frequency, relaxation resonance frequency and 3 dB-frequency as a function of the driving current at 25°C.

All fits show excellent agreement with the measured data. As can be obtained from the Fig. 5.5, the bandwidth increases rapidly with the current and reaches values of larger than 17 GHz already at 2 mA, which corresponds to the nominal current density of only ~7 kA/cm^2. At the current of 3 mA (~11 kA/cm^2) bandwidth becomes larger than 19 GHz and reaches its maximum value of larger than 20 GHz already at 4 mA, which corresponds to the current density of ~14 kA/cm^2. The rapid increase of the modulation bandwidth with the current is an indispensable precondition for high quality optical eye diagrams and thus stable data transmission process at high bit rates. The low operation current density

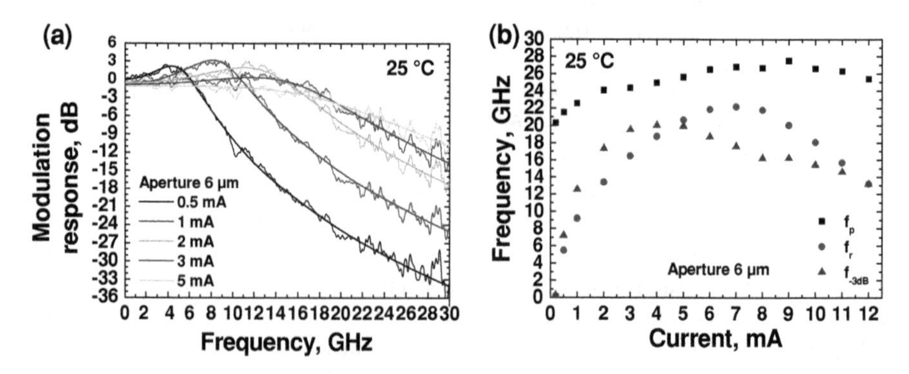

Fig. 5.5 Small signal modulation response with corresponding fits at 25°C for different driving currents (**a**) and extracted values of the parasitic cut-off frequency (*rectangles*), relaxation resonance frequency (*circles*) and 3 dB-frequency (*triangles*) as a function of the current (**b**) for the VCSEL with 6 μm aperture diameter

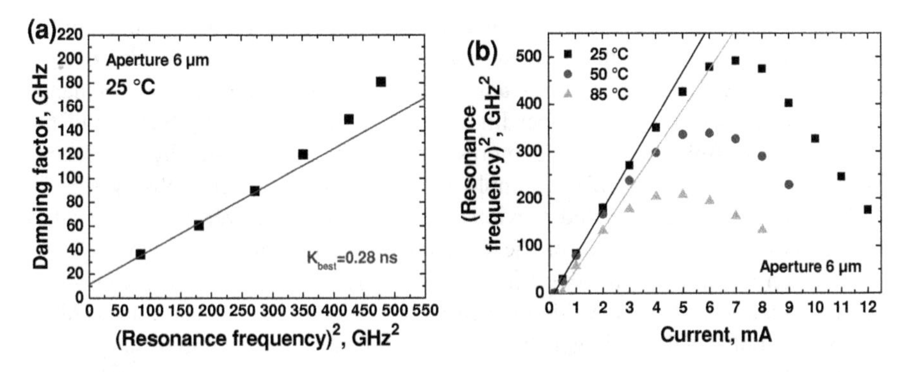

Fig. 5.6 Damping factor as a function of the squared resonance frequency with the corresponding linear fit at 25°C (**a**) and the squared resonance frequency as a function of the driving current at different temperatures with corresponding linear fits (**b**) for the VCSEL with 6 μm aperture diameter

in the range of ~10 kA/cm^2 is one of the major conditions for reliable device operation. Both, the rapid increase of the bandwidth with injected current and the low operation current density, is fulfilled for the investigated 850 nm QW-VCSELs. Small signal modulation bandwidths of 21 and 24 GHz have been demonstrated in the past for VCSELs with oxide-confined apertures of 4 μm in diameter and emitting at 850 [2] and 1100 nm [3], where the latter device is based on a buried tunnel junction. Bandwidths measured with 850 nm QW-VCSELs in this work are very close to these record values.

From the form of the small signal modulation curves shown in Fig. 5.5a one can already recognize that the 6 μm aperture VCSEL is not limited by the thermal effects, because the modulation response curves with bandwidths around the maximum value of ~20 GHz do not exhibit a noticeable relaxation resonance peak. By taking a look on Fig. 5.5b limiting factors for high speed operation of the 6 μm aperture VCSEL can be clearly identified. First, the parasitic cut-off frequency is at all currents at least ~5 GHz larger than the corresponding 3 dB-frequency, and reach values larger than 27 GHz, demonstrating that the limiting effect of the electrical parasitics is noticeable but can not explain the bandwidth limitation to ~20 GHz alone. This increase of the parasitic cut-off frequency was achieved mainly by the further reduction of the parasitic capacitance C_a of the oxide aperture layers and the active region, which is as low as ~128 fF at 5 mA for the 6 μm device.

The relaxation resonance frequency increases also very rapidly with injected current and reaches values larger than 22 GHz, which are also larger than the maximum 3 dB-frequency. Since the relaxation resonance frequency saturates at values larger than the maximum bandwidth, temperature effects do not limit the high speed laser performance. Thus the major contribution to high speed device limitation arises from the combination of electrical parasitics and damping, as can also be obtained from the Fig. 5.6a, where the damping factor as a function of the squared resonance frequency at 25°C is shown.

To extract the K-factor yielding the damping limit of the speed, one must fit the damping as a function of the squared relaxation resonance frequency according to Eq. 2.4.65. The damping factor is not ideally linear with the squared resonance frequency and shows a weak super-linear behavior, most probably because of the heating and possibly due to gain compression, as was also indicated by simulations in the previous chapters (for example in Fig. 2.40). A similar observation in 850 nm VCSELs was reported previously [4], and this observed behavior makes extraction of the exact value of the K-factor difficult. Values between 0.28 and 0.40 ns can be extracted for the 6 μm device, depending on the fitting procedure. For the smallest estimated value of 0.28 ns the corresponding maximum achievable bandwidth would be larger than 30 GHz, for the largest value of 0.40 ns the damping limitation would result in a bandwidth of ~22 GHz, according to (2.4.77). Since at larger resonance frequencies the damping is rather larger, its effect on the modulation bandwidth consequently increases at larger currents. Thus, as already have been mentioned, the 6 μm aperture VCSEL is limited not by the internal heating, but by the combination of the damping and the electrical parasitics.

To investigate high speed properties at elevated temperatures, small signal modulation response curves were measured at three different temperatures: 25, 50 and 85°C. In Fig. 5.6b the corresponding squared relaxation resonance frequencies together with linear fits are shown for the VCSEL with 6 μm aperture diameter. From the linear fits the D-factors at different temperatures can be extracted. At 25°C the D-factor was ~9.9 GHz/Sqrt(mA), which is larger than the D-factors of the corresponding 980 nm SML-VCSELs and 980 nm QW-VCSELs. The measured large D-factor of the 850 nm QW-VCSEL clearly demonstrates the successfully application of the thicker QWs and thus larger optical confinement factor and also of the reduced gain–cavity detuning. The increased optical confinement factor leads to a decrease of the threshold current density and consequently to an increase of the differential gain, which contributes to the increase of the D-factor, according to (2.4.70). The decreased gain peak–cavity resonance detuning helps to increase the gain and the differential gain at 25°C, but has the disadvantage of the decreased temperature stability, as can be obtained from the Fig. 5.6b. The extracted D-factors at 50 and 85°C are ~9.7 and ~9.2 GHz/Sqrt(mA), accordingly, and are lower than the value at 25°C, in opposite to the 980 nm QW-VCSELs, where the D-factor at 50°C had the largest value, which was caused by the larger gain–cavity detuning of nominally 15 nm.

In Fig. 5.7 temperature dependent 3 dB-frequencies and relaxation resonance frequencies at different currents are shown for the 6 μm aperture VCSEL. The maximum modulation bandwidth decreases from ~20 GHz at 25°C to ~19 GHz at 50°C and is still ~16.9 GHz at 85°C. The overall change is about 3 GHz, which is larger than the change of the maximum bandwidth of the 980 nm QW-VCSELs, but it is compensated by the larger value of the maximum bandwidth at 25°C. The 850 nm QW-VCSELs have also at 85°C larger values of the maximum bandwidth, as compared to the 980 nm QW-VCSELs. The reason is the optimized structure the 850 nm QW-VCSEL. The maximum relaxation resonance frequency decreases from ~22 GHz at 25°C to ~18.4 GHz at 50°C, and further

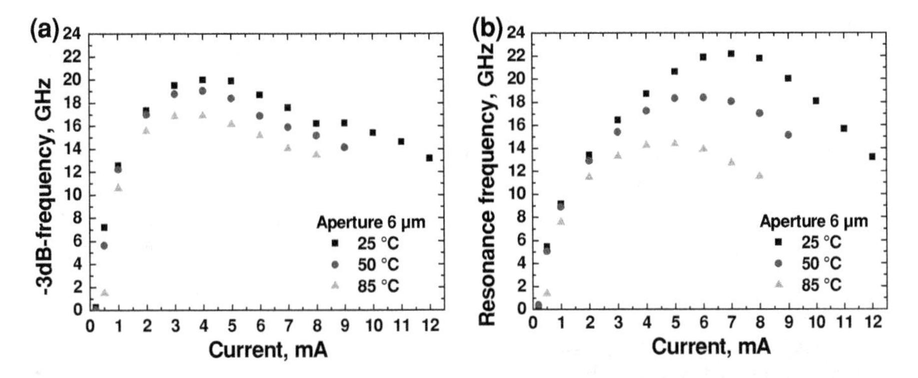

Fig. 5.7 Bandwidth as a function of the driving current (**a**) and relaxation resonance frequency as a function of the driving current (**b**) at different temperatures for the VCSEL with 6 μm aperture diameter

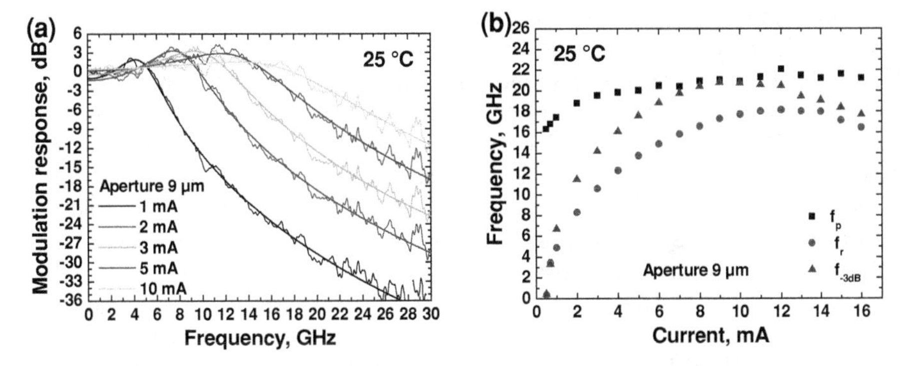

Fig. 5.8 Small signal modulation response with corresponding fits at 25°C for different driving currents (**a**) and extracted values of the parasitic cut-off frequency (*rectangles*), relaxation resonance frequency (*circles*) and 3 dB-frequency (*triangles*) as a function of the current (**b**) for the VCSEL with 9 μm aperture diameter

to ~14.4 GHz at 85°C. At elevated temperatures the 6 μm VCSEL becomes also partially limited by the thermal effects, since the maximum relaxation resonance frequency decreases to values smaller than the 3 dB-frequency.

For the 850 nm QW-VCSEL with a larger oxide aperture diameter of 9 μm the measured small signal modulation response curves at different currents at 25°C with corresponding fits are shown in Fig. 5.8a. In Fig. 5.8b corresponding extracted values of the parasitic cut-off frequency, relaxation resonance frequency and 3 dB-frequency at 25°C are presented.

Maximum 3 dB-frequency larger than 20 GHz has been measured, comparable to the 6 μm aperture VCSEL. The largest value was ~20.8 GHz at driving current of 9 mA, which corresponds to the current density of only ~14 kA/cm^2. Similar to the 6 μm aperture laser, bandwidth of ~17.5 GHz has been achieved at current density of only ~7.8 kA/cm^2. In the 9 μm aperture VCSEL electrical parasitics play a decisive role in the limitation of the high speed laser performance, as can be directly obtained from the Fig. 5.8b, where the extracted maximum parasitic cut-off frequency is with ~22 GHz very close to the measured maximum bandwidth, and at currents between 7 and 11 mA values of both frequencies are practically identical. The reduced parasitic cut-off frequency of the 9 μm aperture VCSEL as compared to the 6 μm aperture device is caused mainly by the larger mesa diameter. Thermal effects also starts to play a noticeable role, but are not dominant, since the maximum extracted relaxation resonance frequency was larger than 18 GHz, which is only ~10% lower than the maximum 3 dB-frequency.

The role of the damping can be estimated by considering the *K*-factor, which can be extracted from the plot of the damping factor as a function of the squared relaxation resonance frequency, as shown in Fig. 5.9a.

From the linear fit the *K*-factor around ~0.21 ns can be extracted, leading to the maximum damping limited bandwidth of larger than 40 GHz, according to (2.4.77). The small value of the *K*-factor confirms the fact, that the damping does not limit the high speed laser performance considerably, and the 9 μm aperture VCSEL is mostly limited by electrical parasitics.

The extracted *D*-factors from the fits of the squared relaxation resonance frequency as a function of the driving current at lower currents, as shown in Fig. 5.9b, are logically lower as compared to the laser with 6 μm aperture diameter, because of the larger mode volume of the 9 μm aperture device. At 25°C the *D*-factor was ~6.5 GHz/Sqrt(mA). It decreased to ~6.4 GHz/Sqrt(mA) at 50°C and further to ~5.7 GHz/Sqrt(mA) at 85°C. As one would expect from the rate equation theory, the *D*-factor should decrease with increasing mode volume, according to Eq. 2.4.70, and due to the square root dependence it is inversely proportional to the diameter of the active region. In fact, dividing the *D*-factor of 9.9 GHz/Sqrt(mA) for the 6 μm VCSEL by the *D*-factor of 6.5 GHz/Sqrt(mA) for the 9 μm VCSEL at 25°C gives the value of 1.52, which is practically identical to the nominal

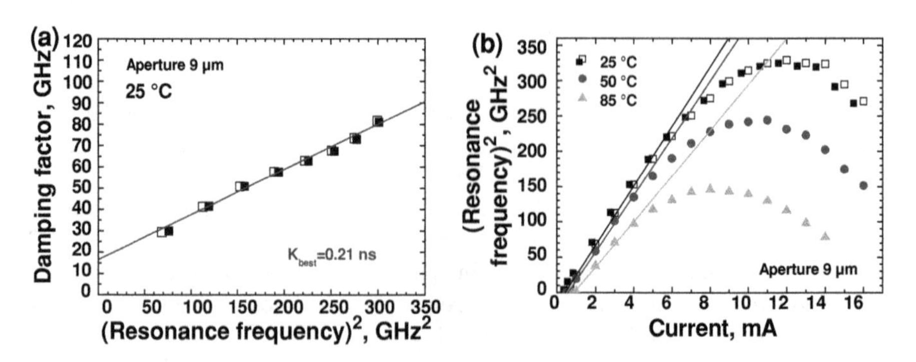

Fig. 5.9 Damping factor as a function of the squared resonance frequency with the corresponding linear fit at 25°C (**a**) and the squared resonance frequency as a function of the driving current at different temperatures with corresponding linear fits (**b**) for the VCSEL with 9 μm aperture diameter

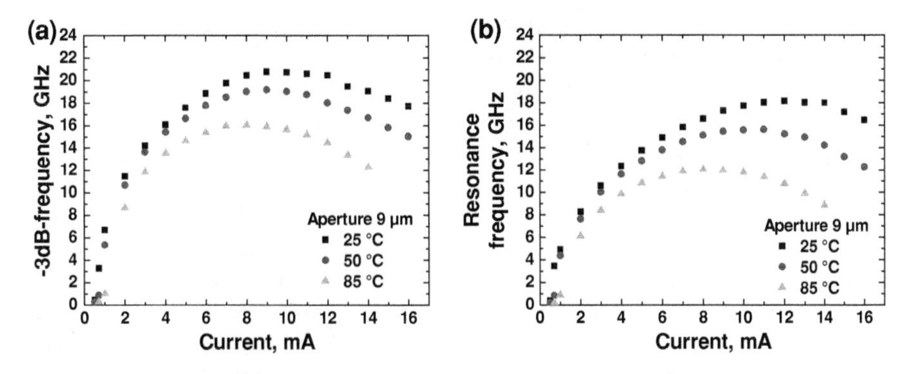

Fig. 5.10 Bandwidth as a function of the driving current (**a**) and relaxation resonance frequency as a function of the driving current (**b**) at different temperatures for the VCSEL with 9 μm aperture diameter

aperture diameter ratio of 9:6 = 1.5. Clearly our analysis is consistent not only for one particular device but for the entire set of devices, confirming the high quality of the growth and fabrication processes.

At elevated temperatures both the 3 dB-frequency and the relaxation resonance frequency decreases, as shown in Fig. 5.10 for the 9 μm aperture VCSEL. The maximum bandwidth decreases from ~20.8 GHz at 25°C to ~19.2 GHz at 50°C, and is still larger than 16 GHz at 85°C. These values are comparable to the values of the 6 μm aperture device.

The maximum relaxation resonance frequency demonstrates a similar behavior and decreases from ~18 GHz at 25°C to ~15.6 GHz at 50°C and further to ~12 GHz at 85°C, showing an increased impact of thermal effects at elevated temperatures. At the temperature of 85°C the maximum bandwidth of the investigated 850 nm QW-VCSELs is still larger than the corresponding values of the 980 nm QW-VCSELs, indicating at least similar high speed performance of the 850 nm QW-VCSELs at this temperature.

Summarizing, the performed small signal modulation experiments on the 850 nm QW-VCSELs have demonstrated very high modulation bandwidths of larger than 20 GHz for devices with aperture diameters of 6 and 9 μm at current densities of only ~14 kA/cm^2, which is very important for the laser stability and reliability. Relaxation resonance frequencies larger than 22 GHz have been measured. Also at elevated temperature of 85°C the measured maximum bandwidth was larger than 16 GHz. Achieved results show a very high potential of the investigated 850 nm QW-VCSELs for high speed data transmission with bit rates of 35 Gbit/s and larger, which will be confirmed by the large signal modulation experiments and bit error rate measurements presented in the next section.

Optimized device structure and design have lead to the reduction of electrical parasitics, resulting in the measured maximum parasitic cut-off frequencies larger than 27 GHz for the 6 μm aperture laser. Because of the larger mesa diameter, for the 9 μm aperture device the maximum parasitic cut-off frequency was ~22 GHz, limiting the high speed laser performance in a strong way. For future VCSELs aiming bit rates larger 40 Gbit/s further reduction of electrical parasitics is indispensable and can be relatively easy realized by the further reduction of the mesa diameter for the identical diameter of the oxide aperture and/or by introducing additional layers with high Al content for forming additional oxide apertures during the wet oxidation process. The latter variant has the disadvantage of the introduction of additional mechanical stress into the VCSEL structure, which could have negative effect on the laser performance and reliability, and therefore should be applied more carefully. In opposite, the reduction of the mesa diameter can be realized relatively straightforward, since in this work the smallest mesa diameters were nominally 25 μm, and this value could be further reduced to 20 or even 18 μm without sacrificing the simplicity of the fabrication process and without having a noticeable negative effect on the thermal properties of the VCSELs. Additionally to electrical parasitics, damping has played a certain role in the high speed laser limitation for smaller devices. For the high speed high temperature operation also management of the thermal effects should be optimized. Both can be achieved by the further optimization of the laser structure and will be the subject of the future work on high speed VCSELs.

5.4 Large Signal Modulation Characteristics

To investigate the laser behavior in the data transmission process, large signal modulation experiments were performed using a non-return to zero (NRZ) data pattern in the back-to-back configuration (BTB, $\sim$3 m-length fiber) with a $2^7 - 1$ pseudorandom bit sequence (PRBS) from an SHF 12100B bit pattern generator (rise time of 8 ps). The NRZ signal was amplified by a 40 GHz amplifier (+20 dB; rise time of 6 ps) and then attenuated by fixed 60 GHz attenuators to achieve the required amplitude. The RF-signal was combined with a DC bias through a 60 GHz bias-T and fed to on-wafer VCSELs by a high frequency GSG-probe head. The VCSELs were butt-coupled to a 62.5 μm-core multimode fiber (with an estimated 50% coupling efficiency) connected to a JDSU OLA-54 variable optical attenuator, after which the optical signal was sent to the photodetector coupled with either a 38 (+26 dB, rise time of 8 ps) or 40 GHz (+17 dB, rise time of 8 ps) amplifier via a short patch cord. The photodetector chosen was an FC-connectorized multimode fiber pigtailed module from VI Systems GmbH (model D30-850 M) with a rise time (20–80%) below 7 ps. The average power of the detected signal was monitored by a JDSU OLP-55 power meter. Optical eye diagrams and bit error rate curves were studied using a 70 GHz Agilent 86100C digital oscilloscope, combined with a precision time base and an SHF 11100B bit error analyzer, respectively.

Optical eye diagrams were measured at room temperature for different bit rates up to 40 Gbit/s for the VCSEL with the aperture diameter of 6 μm and for the VCSEL with the aperture diameter of 9 μm. In Fig. 5.11 the optical eye diagrams of the 6 μm VCSEL measured at the driving current of 5 mA and peak-to-peak

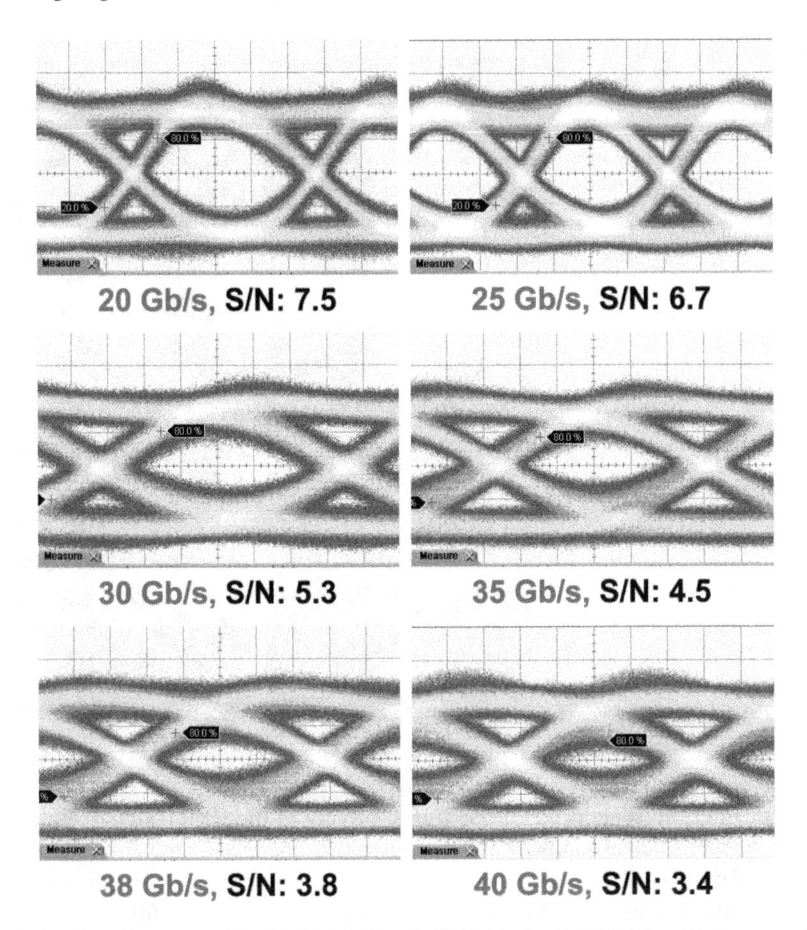

Fig. 5.11 Eye diagrams at 20, 25, 30, 35, 38 and 40 Gbit/s for the VCSEL with 6 μm aperture diameter at operation current of 5 mA and peak-to-peak modulation voltage $V_{p\text{–}p}$ of 0.8 V in a BTB configuration with $(2^7 - 1)$ PRBS, NRZ and corresponding signal-to-noise ratios (S/N) at room temperature

modulation voltage $V_{p\text{–}p}$ of 0.8 V with corresponding bit rates and signal-to-noise ratios (SNRs) are presented. Up to bit rates of 35 Gbit/s all eyes are clearly open with signal-to-noise ratio large than 4, and at bit rates of 20, 25 and 30 Gbit/s the signal-to-noise ratio is larger than 5. From the bit rate of 38 Gbit/s the eyes begin to close, but even at 40 Gbit/s some opening could be obtained. To ensure an error free operation in the cases, where the eyes are not clearly open, direct bit error rate measurements should be carried out. BER measurements have been performed for different devices and are presented below for the VCSEL with the aperture diameter of 9 μm, since this device has demonstrated the larger eye opening. It should also be noted that the eye quality of the received signal for all measured optical eye diagrams is noticeable affected by the imperfection of the electrical RF-signal applied to the devices (thus one must account for measurement system

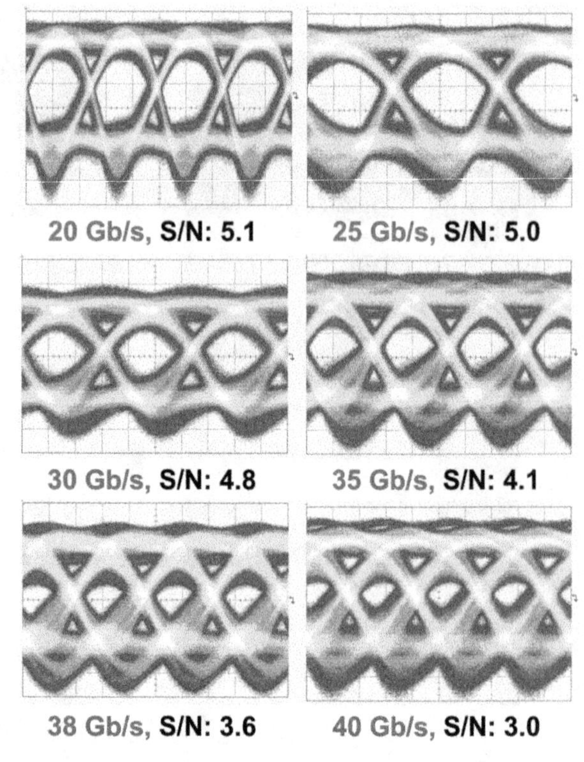

Fig. 5.12 Eye diagrams at 20, 25, 30, 35, 38 and 40 Gbit/s for the VCSEL with 9 μm aperture diameter at operation current of 9 mA and peak-to-peak modulation voltage V_{p-p} of 0.8 V in a BTB configuration with $(2^7 - 1)$ PRBS, NRZ and corresponding signal-to-noise ratios (S/N) at room temperature

errors of jitter, noise, additional rise times, ringing in the BERT signal, etc.). The current of 5 mA corresponds for the 6 μm nominal aperture diameter to the current density of only ~18 kA/cm^2, close to the value of 10 kA/cm^2, at which common commercial available 10 Gbit/s VCSELs operate, and which is sufficiently low to ensure reliable laser operation.

In Fig. 5.12 optical eye diagrams at different bit rates with corresponding signal-to-noise ratios are shown for the VCSEL with 9 μm aperture diameter at the driving current of 9 mA and peak-to-peak modulation voltage V_{p-p} of 0.8 V. The eyes for the 9 μm VCSEL have lower SNRs than for the 6 μm VCSEL, among other because of the presence of a larger number of the lasing modes and corresponding mode competition noise for the larger aperture VCSEL. Nevertheless the eye opening for the 9 μm device is larger as compared to the 6 μm aperture VCSEL. The eyes are clearly open up to bit rates of 35 Gbit/s with the corresponding signal-to-noise ratios large than 4. At bit rates of 20 and 25 Gbit/s the signal-to-noise ratio is equal to or larger than 5. Because of the larger output power modulation, the eye opening is more distinct also at higher bit rates of 38 and 40 Gbit/s. The current of 9 mA corresponds for the 9 μm nominal aperture diameter to the current density of only ~14 kA/cm^2, which is again very close to the value of 10 kA/cm^2.

To ensure an error free operation of the investigated VCSELs, bit error rates at different bit rates were measured. The highest bit rate with the BER below 10^{-12}

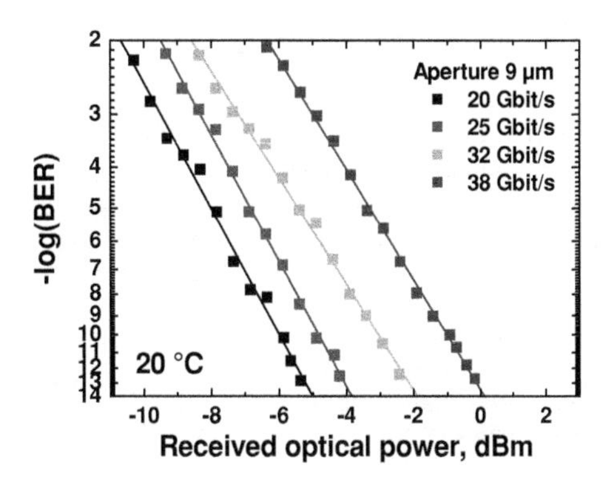

Fig. 5.13 BER measurements in a BTB configuration with $(2^7 - 1)$ PRBS, NRZ for the 9 μm aperture VCSEL at bit rates of 20, 25, 32 and 38 Gbit/s for the bias current of 9 mA and peak-to-peak modulation voltage of 0.8 V at the temperature of 20°C with corresponding linear fits

and no error floor was 38 Gbit/s for the 9 μm VCSEL at the temperature of 20°C, as can be obtained from Fig. 5.13. The peak-to-peak modulation amplitude $V_{p\text{–}p}$ was 0.8 V, while the bias current was fixed at 9 mA, both values identical to the values used for the eye diagram measurements. The detector saturation at higher average powers and internal amplifier noise prevented among other robust error free transmission at 40 Gbit/s. VCSELs with smaller aperture diameters have demonstrated lower error free bit rates, because of the lower output power and accordingly smaller eye opening. The performed BER measurements have demonstrated the error free data transmission at the record high bit rate of 38 Gbit/s at sufficiently low current density at the important wavelength of 850 nm. This was the worldwide first demonstration of the error free data transmission with an 850 VCSEL at 38 Gbit/s. This was also the worldwide first demonstration of the error free operation at 38 Gbit/s using an oxide-confined VCSEL at any wavelength, which is a very important step on the way to the future VCSEL-based optical communication systems, since oxide-confined VCSELs have established themselves as very reliable and cheap laser sources and dominate the VCSEL market today.

Although the 850 nm QW-VCSELs were optimized to operate at room temperature, because of the success of the applied improvement concepts, resulted in the record high speed characteristics, also high temperature laser performance was improved. Fig. 5.14 demonstrates BTB (~3 m multimode fiber), NRZ, $2^7 - 1$ PRBS optical eye diagrams for the 850 nm QW-VCSEL with the aperture diameter of 7 μm at bit rates of 20 and 25 Gbit/s and at two different temperatures: 25 and 85°C. The eyes are clearly open at 25°C and also at the elevated temperature of 85°C, with corresponding signal-to-noise ratios larger than 6 for the room temperature eye diagrams and larger than 4 for the eye diagrams measured at 85°C. These values are very similar to the signal-to-noise ratios measured for the 980 nm QW-VCSEL with nominally the same aperture diameter of 7 μm presented in the previous chapter. Also the optical eye diagrams are comparable. The reason is the similar bandwidth of both 850 and 980 nm QW-VCSELs at 85°C, enabling the 850 nm QW-VCSELs to operate at this elevated temperature at the bit rates up to 25 Gbit/s. Thus, the improved VCSEL structure and device design have compensated the decreased gain peak–cavity resonance detuning of the 850 nm

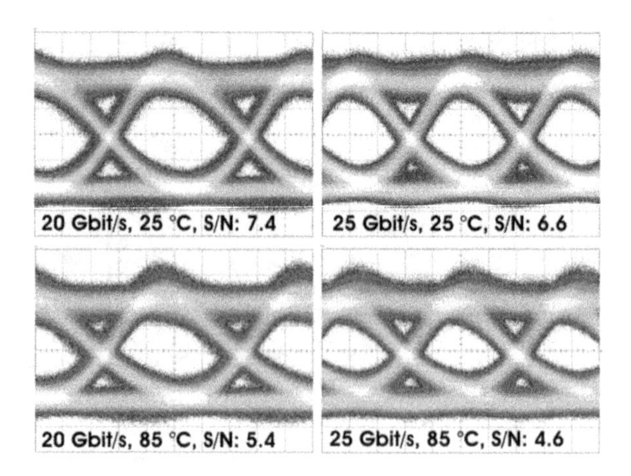

Fig. 5.14 Eye diagrams at temperatures of 25 and 85°C and bit rates of 20 and 25 Gbit/s for the VCSEL with 7 μm aperture diameter at operation currents of 5.5 mA at 25°C and 6.2 mA at 85°C with peak-to-peak modulation voltage $V_{p\text{–}p}$ of 0.8 V in a BTB configuration with $(2^7 - 1)$ PRBS, NRZ and corresponding signal-to-noise ratios (S/N)

QW-VCSELs as compared to the 980 QW-VCSELs, leading to the practically identical high temperature performance. Data transmission with the bit rate of 25 Gbit/s at 85°C has been recently demonstrated [1] using 850 nm oxide-confined VCSELs. Thus our results of the 850 nm QW-VCSELs have reproduced this worldwide record achievement at 85°C, demonstrating at the same time the record high data transmission bit rate of 38 Gbit/s at room temperature, compared to the highest previously achieved bit rate of 32 Gbit/s at 850 nm [1], confirming the importance of the applied optimization concepts both for the VCSEL epitaxial structure and also for the device design.

To ensure an error free data transmission also at elevated temperatures, BER measurements at bit rates of 20 and 25 Gbit/s and temperatures of 25 and 85°C have been carried out for the 7 μm aperture 850 nm QW-VCSEL, directly corresponding to the bit rates and temperatures used for the optical eye diagrams measurements. The results are presented in Fig. 5.15.

The investigated 7 μm aperture VCSEL operated error free at bit rates up to 25 Gbit/s at 85°C, as could be already expected by considering the optical eye diagrams at the corresponding temperatures and bit rates presented previously in Fig. 5.14.

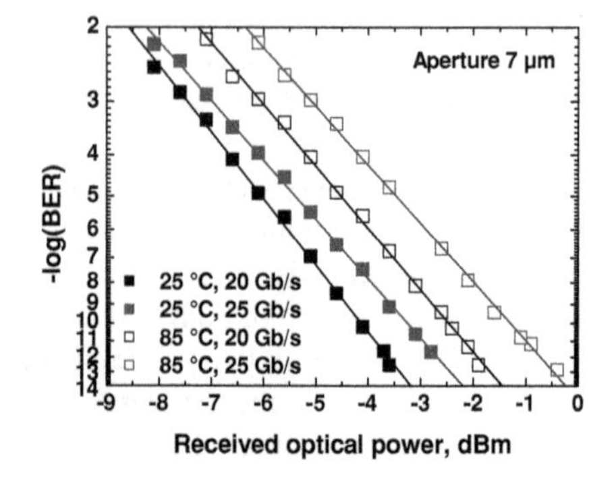

Fig. 5.15 BER measurements in a BTB configuration with $(2^7 - 1)$ PRBS, NRZ for the 7 μm aperture VCSEL at temperatures of 25 and 85°C and bit rates of 20 and 25 Gbit/s at operation currents of 5.5 mA at 25°C and 6.2 mA at 85°C and peak-to-peak modulation voltage of 0.8 V

The large signal modulation and BER measurement results of the 850 nm QW-VCSELs presented in this section demonstrated error free data transmission at the record high bit rate of 38 Gbit/s. Additionally, error free operation at 25 Gbit/s at 85°C was demonstrated. These application oriented measurements have confirmed improvement of the physical laser properties by applied optimization concepts.

5.5 Summary of the 850 nm QW-VCSEL Results

To summarize, worldwide first VCSELs demonstrating error free data transmission at the record high bit rate of 38 Gbit/s at the commercial important wavelength of 850 nm were realized. The operation current density was only $\sim$14 kA/cm^2, sufficiently low to ensure reliable and stable device operation. Also error free operation at the elevated temperature of 85°C at the bit rate of 25 Gbit/s was demonstrated.

The reason for this remarkable high speed performance was the consequently large small signal modulation bandwidth of larger than 20 GHz for VCSELs with different aperture diameters. To achieve such large bandwidths, several optimization concepts have been consequently applied to overcome physical limitations for high speed laser operation. By introducing two oxide apertures, reducing VCSEL mesa diameters, applying thick dielectric layers, reducing metal contact areas, utilizing high frequency impedance matched GSG contact pads and carefully optimizing the doping levels in the whole laser structure, the limiting effects of the electrical parasitics and of the disturbing reflections of high frequency electrical signals have been efficiently suppressed, leading to cut-off frequencies of electrical parasitics of larger than 27 GHz for smaller devices. To overcome the thermal limitation improved epitaxial laser structure and the optimized double mesa device design were applied. These improvements, together with the application of the advanced active region with the larger differential gain, have lead to the saturation of the relaxation resonance frequency at values of larger than 22 GHz for smaller devices and also to a rapid increase of the relaxation resonance frequency and of the bandwidth with the injected current. Logically, in VCSELs with larger aperture diameters both electrical parasitics and thermal effects have played a noticeable role, but nevertheless also in the larger aperture devices bandwidths of larger than 20 GHz have been achieved. This was possible because of the significant reduction of the damping, arising from the optimization of the cavity and of the active region, leading to the damping limited bandwidths of larger than 40 GHz for larger devices.

Decisive advantages of the applied optimization concepts, beside improvement of the high speed laser performance, are the relative overall simplicity of the VCSEL manufacturing process, the absence of the complicated and thus expensive steps, the straightforward device growth and fabrication, all of them increasing the suitability of the MOCVD-grown 850 nm QW-VCSELs presented in this chapter for the commercial large-scale cheap mass production. The low operation current density and the absence of the critical growth and fabrication steps improve the laser stability and reliability, making the real world application of such 850 nm QW-VCSELs feasible.

References

1. Westbergh P, Gustavsson JS, Haglund A, Larsson A, Hopfer F, Fiol G, Bimberg D, Joel A (2009) 32 Gbit/s multimode fiber transmission using high-speed, low current density 850 nm VCSEL. IEEE Electron Lett 45(7)
2. Lear KL, Hietala VM, Hou HQ, Ochiai M, Banas JJ, Hammons BE, Zolper J, Kilcoyne SP (1997) Small and large signal modulation of 850 nm oxide-confined vertical cavity surface emitting lasers. In: Advances in vertical cavity surface emitting lasers in series OSA trends in optics and photonics, vol 15, pp 69–74
3. Anan T, Suzuki N, Yashiki K, Fukatsu K, Hatakeyama H, Akagawa T, Tokutome K, Tsuji M (2007) High-speed InGaAs VCSELs for optical interconnections. In: International symposium on VCSELs and integrated photonics, Tokyo, Japan, 17–18 December 2007, pp 76–78
4. Westbergh P, Gustavsson JS, Haglund A, Sköld M, Joel A, Larsson A (2009) High-speed, low-current-density 850 nm VCSELs. IEEE J Sel Top Quant Electron 15(3):694–703

Chapter 6
Conclusions and Outlook

6.1 Summary

The main goal of the present work was to develop high speed readily manufacturable GaAs-based oxide-confined VCSELs emitting around 850 and 980 nm for future applications in optical networks, both outside and inside computers. While at the wavelength of 850 nm the main goal was to maximize the bit rate at room temperature, for devices emitting at 980 nm an additional requirement was to achieve high temperature stability. Special attention was paid to development of a straightforward manufacturing process, which would not require complicated growth or fabrication steps, making the developed VCSELs suitable for inexpensive large-scale industrial mass production.

To characterize and consequently improve high speed laser performance advanced characterization techniques were applied, including small signal modulation response (S_{21}) and electrical microwave reflection (S_{11}) measurements, large signal modulation experiments (optical eye diagrams) and bit error rate (BER) measurements. VCSEL behavior at different temperatures from 0 up to 120°C was investigated, enabling comprehensive device characterization.

Within the scope of this dissertation three generations of VCSELs were developed and investigated. Results of the former generations were utilized to improve lasers from the latter generations. The first two generations were developed to emit laser light at 980 nm and have different types of active regions inside: InGaAs layers grown in the submonolayer growth mode, applied in the 980 nm SML-VCSELs, and InGaAs quantum wells, utilized in the 980 nm QW-VCSELs. The VCSELs from the third generation were devices emitting at 850 nm, with InGaAs quantum wells in the active region.

To combine high temperature stability with the proper high speed performance advanced concepts were applied in the 980 nm SML-VCSELs. Doping levels and layer compositions and thicknesses within both DBR mirrors and the cavity region were optimized to achieve high thermal conductivity, low electrical resistance and low optical losses. Highly strained InGaAs active layers grown in the submonolayer

A. Mutig, *High Speed VCSELs for Optical Interconnects*, Springer Theses,
DOI: 10.1007/978-3-642-16570-2_6, © Springer-Verlag Berlin Heidelberg 2011

growth mode together with the gain peak–cavity resonance detuning of 15 nm have decisively contributed to the increased temperature stability of the devices. Applied optimization concepts resulted in realization of worldwide first VCSELs operating at 20 Gbit/s at elevated temperatures of up to 120°C. This is at the moment the highest operation temperature at 20 Gbit/s ever reported for any VCSEL. Very temperature insensitive CW and HF characteristics have been measured. The hardly temperature dependent maximum bandwidth was obtained, which changed by less than 2 GHz from 12.9 GHz at 25°C to 11 GHz at 120°C. As main limiting factors electrical parasitics and damping were recognized. At larger aperture diameters also saturation of the relaxation resonance frequency caused by thermal effects was considerable. Consequently, further optimizations of the laser structure and device design were necessary to overcome these limitations, which have led to realization of the 980 nm QW-VCSELs.

By reducing laser mesa diameters, introducing double oxide aperture and increasing the BCB thickness, electrical parasitic cut-off frequency could be drastically increased to values of larger than 22 GHz, enabling to overcome the corresponding limitation in the 980 nm QW-VCSELs. To increase relaxation resonance frequencies, especially for devices with larger aperture diameters, optimized active region based on compressively strained InGaAs QWs with tensile strained GaAsP barriers has been introduced. This has enabled to increase the D-factors by more than a factor of two compared to the 980 nm SML-VCSELs with the same aperture diameters. These improvements together with the gain peak–cavity resonance detuning of 15 nm, nominally identical to the detuning of 980 nm SML-VCSELs, resulted in an increase of the maximum bandwidth by more than 3 GHz at 25°C, while maintaining the high temperature stability of both CW and HF characteristics. Clearly open optical eye diagrams at bit rates of up to 30 Gbit/s at 25°C and up to 25 Gbit/s at 85°C have been obtained, increasing the maximum achieved high temperature bit rate by 5 Gbit/s as compared to the 980 nm SML-VCSELs. Small signal modulation measurements have localized damping as the major limitation of the high speed performance of the 980 nm QW-VCSELs. Consequently, to increase the bit rate further, additionally concepts were applied in the third generation of VCSELs to efficiently reduce the damping and to increase both the cut-off frequency of electrical parasitics and the relaxation resonance frequency.

The 850 nm QW-VCSELs were optimized to achieve the maximum possible bit rate at room temperature, since the main application field of the GaAs-based oxide-confined 850 nm VCSELs is the data transmission in LAN and SAN outside computers. Thus, reduced gain peak–cavity resonance detuning of nominally 8 nm was applied, leading to a decrease of the threshold carrier density and consequently to an increase of the differential gain and of the relaxation resonance frequency at room temperature. Further optimizations of the active region based on compressively strained InGaAs QWs and of the cavity region resulted in increased relaxation resonance frequencies. Values larger than 22 GHz have been measured. Further reduction of the mesa diameters for the same aperture diameters and application of the double oxide aperture led to a decisive increase of the electrical parasitic cut-off frequency to values larger than 27 GHz. By optimizing among other the mirror reflectivities, damping was sufficiently reduced, leading to the maximum achievable internal damping limited bandwidth of larger than 40 GHz. These optimizations resulted in record values of the maximum bandwidth of larger than 20 GHz at low current densities of only ~14 kA/cm^2, sufficiently low to ensure reliable device operation. Consequently, error free data transmission at bit rates as large as 38 Gbit/s was achieved at room temperature. These were worldwide first 850 nm VCSELs operating at such high bit rates. These were also worldwide first of any oxide-confined VCSELs demonstrating error free operation at 38 Gbit/s. The increase of the maximum demonstrated data transmission bit rate of 850 nm VCSELs from previously demonstrated 32 Gbit/s by the group of Prof. Dr. Larsson [1, 2] to 38 Gbit/s, achieved within the scope of this dissertation, is an important step toward 40 Gbit/s, which clearly demonstrates the high potential and suitability of the GaAs-based oxide-confined VCSELs for upcoming ultra-high speed applications at 40 Gbit/s and beyond.

To conclude and summarize, the research and development work carried out within the scope of the present dissertation has decisively contributed to increase of both maximum data transmission bit rate and operation temperature of GaAs-based oxide-confined VCSELs and has moved these devices one step closer to practical applications in the future ultra-high speed short reach optical interconnects.

6.2 Future Works

The research and development activities performed within the scope of this dissertation are one of the first steps on the way towards reliable, readily manufacturable, ultra-high speed VCSELs for future optical interconnects operating at bit rates far above 40 Gbit/s. Although many advanced concepts were applied for laser investigated here, the room for further improvements is still large, increasing the chances for success. As was demonstrated in the previous chapters, high speed VCSEL characteristics are commonly affected by two or even all three types of limitations: electrical parasitics, thermal effects and damping. Consequently, to improve the high speed laser performance, especially at elevated temperatures, further optimizations are indispensable, which will overcome all types of limitations.

The rules to increase the cut-off frequency of electrical parasitics are pretty straightforward: all parasitic capacitances and resistances within the device should be reduced. The main challenge is to find the proper compromise between the concepts for reducing electrical parasitic elements and other applied optimization concepts, since they interact and affect each other. The common concept to reduce electrical resistances inside the device is to optimize the doping levels in the structure. The main trade-off here is to keep optical absorption losses at an acceptable level, since the free carrier absorption losses increase with increasing doping levels. Also further optimizations of the annealing process of the metal contacts could lead to a noticeable reduction of the total device resistance [3]. Additionally, VCSELs grown on p-doped substrates can be utilized, where the largest part of the thick bottom mirror is kept undoped and the top mirror becomes n-doped. Since the resistance of the n-doped AlGaAs is lower than that of the p-doped material, the overall resistance of the device can be lowered. Hereby an intracavity p-contact is required, making the carefully optimization of the epitaxial structure and fabrication process very critical.

Generally, to reduce parasitic capacitances VCSEL dimensions should be reduced and corresponding thicknesses of epitaxial and dielectric layers should be increased. The main trade-off by the reduction of the VCSEL mesa diameter is to maintain a reasonable good thermal conductivity, since it reduces with the reduction of the laser area. The fact that the thermal conductivity reduces linear with the aperture diameter and thus with the mesa diameter for oxide-confined VCSELs, as given by (2.3.10), while capacitances decrease with the square of the mesa diameter, facilitates the task. Additionally, device design with two mesas can be applied, as was the fact for the VCSELs described in this work. Reduction of the mesa diameter to 18 μm or even less without decisive increase of the laser thermal resistance appears to be realistic.

Increasing the thickness of the dielectric layers, which reduces the parasitic capacitance of the contact pads, does not appear to have any negative trade-offs excepted possible increase of the laser fabrication effort. The maximum thickness here is mostly given by the VCSEL height. Reduction of the capacitance of the laser mesa itself, which includes among other capacitance of the oxide layers, can be achieved by introduction of several deep oxidation layers [4, 5] additionally to the double oxide aperture utilized in the QW-VCSELs described in this dissertation. These are layers in the DBR mirror, in which the Al composition is lower than in the aperture layers, but higher than in the mirror layers. Thus their oxidation lengths can be calibrated to be e.g. a half of the oxidation length of the aperture layers, effectively increasing the thickness of the oxidized layers and thus lowering the corresponding parasitic capacitance. The epitaxial effort is practically identical compared to a conventional VCSEL structure, while significant reduction of the mesa capacitance can be achieved. The main trade-off here is the increase of the electrical resistance, so that carefully optimization of the oxidation depth and number of the deep oxidation layers should be performed. To reduce mesa capacitance proton implantation can be applied as well, which however requires an additional fabrication step [6].

To increase the maximum relaxation resonance frequency and thus to overcome the thermal limitation different concepts can be utilized. First, the active region can be optimized further in order to increase the differential gain, which will immediately increase the relaxation oscillation frequency, according to (2.4.70). This can be done by the further theoretical and experimental investigations of the effect of the In composition in the QWs, QW thickness, and barrier thickness and composition on the differential gain [7]. Also advanced active region concepts, for example quantum dots (QD), are possible [8–10]. As can be obtained from (2.4.70), reduction of the mode volume will result in an increase of the D-factor, leading to larger relaxation resonance frequency. Since decrease of the oxide aperture diameter to smaller values has a strong negative effect on the device performance via among other increased resistance, increased current density, larger thermal resistance and decreases device reliability, other concepts, for example photonic crystal VCSELs [11, 12], are more preferable. Also further optimizations of the cavity region and of the overall laser structure can lead via e.g. decreased losses, and thus decreased threshold carrier densities, to an increase of the differential gain, accordingly to (2.4.23) and (2.4.36). Since the photon density has an effect on the differential gain via gain compression factor, proper balance should be found while optimizing the VCSEL cavity and mirror reflectivities.

Another side of the optimizations aiming increased relaxation resonance frequencies is the improvement of the thermal conductivity of the VCSELs. This will enable to reach larger driving currents before thermal roll-over, and thus to increase the maximum bandwidth, according to (2.4.70). For this purpose e.g. advanced heat sink techniques could be utilized [13]. A relatively straightforward way is also to use Al composition of 100% in the largest part of the bottom mirror, and in the case of 980 nm VCSELs to use binary AlAs/GaAs bottom mirror. Since the thermal conductivity of the binary materials are much better compared to ternary alloys, significant increase of the thermal conductivity can be achieved, increasing the roll-over current. Carefully in situ controlling of the etching depth is in the case of binary bottom mirrors indispensable to avoid undesirable mirror oxidation in the following oxidation process. An additional advantage of binary mirrors is the reduced effective cavity length, which increases the optical confinement factor and thus decreases the threshold carrier density. Additionally, the shorter cavity length reduces photon life time, helping to decrease damping, which is an indispensable precondition for proper high speed laser operation.

Accordingly to (2.4.67), efficiently reduction of the damping in the structure can be achieved by the reduction of the photon life time and of the gain compression factor and by the increase of the differential gain. Additionally to the reduction of the effective cavity length and thus photon life time by application of binary mirrors, shorter cavity regions can be utilized. Since in this dissertation all VCSELs have the optical length of the cavity region equal to 3/2 λ, the next step would be to reduce the optical length of the cavity to $\lambda/2$. This would considerably increase the maximum achievable damping limited bandwidth, without significant direct negative effects on other important laser parameters. Also carefully optimization of the mirror reflectivities gives access to adjustment of the photon life time. However, changing the mirror reflectivities affects also other decisive laser characteristics, among other losses, threshold carrier density, differential gain, etc. Thus, comprehensive studies should be carried out, both theoretically and experimentally, in order to find the optimum for the best VCSEL high speed performance. To reduce the gain compression factor and to increase the differential gain, optimizations of the active region are required. Also the cavity design has effect on the gain compression. Precise control of the laser modal properties can additionally help to improve the VCSEL properties.

As can be obtained from the described possibilities for further improvements of the VCSELs developed in this dissertation, difficult but at the same time very interesting and exciting work is waiting to be done. On this way many challenges have to be overcome and many problems have to be solved. While many described concepts are relatively straightforward by their self, the main challenge is to combine them and to find the proper compromise between the different competitive trends. Also novel concepts for indirect modulation of the VCSEL output power, e.g. electro-optically modulated VCSELs (EOM-VCSELs) [14–16], are very promising and will be investigated further. In addition to the actual laser development, new concepts for the advanced VCSEL packaging, mounting and exact positioning should be elaborated. This is especially important for chip-to-chip and on-chip optical interconnects, where the main trend is the vertical integration of chips and interconnects [17]. For integration of VCSELs to CMOS chips novel concepts, e.g. lift-off VCSELs [18, 19], should be intensively investigated. The reward will be efficient, low power, reliable, readily positionable VCSELs operating at ultra-high speed, enabling realization of the next generations of optical interconnects.

Of course, the field of photonics and optoelectronics is not limited to VCSELs. In the long-term perspective revolutionary photonic devices [20–23] will come to our daily life, reducing the dimensions and power consumption and increasing the bandwidth of optical interconnects by several orders of magnitude. These novel nanolasers, based among other on scientific results received from the VCSEL research, will bring the human society one step further on the way of the non-stopping progress.

References

1. Westbergh P, Gustavsson JS, Haglund A, Sköld M, Joel A, Larsson A (2009) High-speed, low-current-density 850 nm VCSELs. IEEE J Sel Top Quant Electron 15(3):694–703
2. Westbergh P, Gustavsson JS, Haglund A, Larsson A, Hopfer F, Fiol G, Bimberg D, Joel A (2009) 32 Gbit/s multimode fiber transmission using high-speed, low current density 850 nm VCSEL. IEEE Electron Lett 45(7):366–368
3. Anuar MSK, Sharizal AM, Mitani SM, Razman YM, Mat AFA, Choudhury PK (2006) Effect of rapid thermal annealing (RTA) on n-contact of 980 nm oxide VCSEL. In: Proceedings of the IEEE international conference on semiconductor electronics ICSE 2006, Kuala Lumpur, Malaysia
4. Ou Y, Gustavsson JS, Westbergh P, Haglund A, Larsson A, Joel A (2009) Impedance characteristics and parasitic speed limitations of high speed 850 nm VCSELs. IEEE Photonics Technol Lett 21(24):1840–1842
5. Chang Y-C, Wang CS, Coldren LA (2007) Small-dimensions power-efficient high-speed vertical-cavity surface-emitting lasers. Electron Lett 43(7):396–397
6. Cheng YM, Herrick RW, Petroff PM, Hibbs-Brenner MK, Morgan RA (1995) Degradation studies of proton-implanted vertical cavity surface emitting lasers. Appl Phys Lett 67(12):1648–1650
7. Chang Y-A, Chen J-R, Kuo H-C, Kuo Y-K, Wang S-C (2006) Theoretical and experimental analysis on InAlGaAs/AlGaAs active region of 850-nm vertical-cavity surface-emitting lasers. J Lighwave Technol 24(1):536–543

8. Bimberg D (2008) Quantum dot based nanophotonics and nanoelectronics. IEEE Electron Lett 44(33):168–171
9. Krestnikov IL, Ledentsov NN, Hoffmann A, Bimberg D (2008) Arrays of two-dimensional islands formed by submonolayer insertions: growth, properties, devices. Phys Stat Sol (a) 183(2):207–233
10. Bimberg D, Grundmann M, Ledentsov NN (1999) Quantum dot heterostructures. Wiley, Chichester
11. Danner AJ, Raftery JJ Jr, Leisher PO, Choquette KD (2006) Single mode photonic crystal vertical cavity lasers. Appl Phys Lett 88:091114
12. Leisher PO, Danner AJ, Raftery JJ Jr, Siriani D, Choquette KD (2006) Loss and index guiding in single-mode proton-implanted holey vertical-cavity surface-emitting lasers. IEEE J Quant Electron 42(10):1091–1096
13. AL-Omari AN, Carey GP, Hallstein S, Watson JP, Dang G, Lear KL (2006) Low thermal resistance, high speed, top emitting 980 nm VCSELs. IEEE Photonics Technol Lett 18(11):1225–1227
14. Shchukin VA, Ledentsov NN, Lott JA, Quast H, Hopfer F, Karachinsky LYa, Kuntz M, Moser P, Mutig A, Strittmatter A, Kalosha VP, Bimberg D (2008) Ultra high-speed electro-optically modulated VCSELs: modeling and experimental results. Physics and simulation of optoelectronic devices XVI, Invited paper, Proceedings of the SPIE, vol 6889, 68890H, 22 February 2008
15. Paraskevopoulos A (2006) High-bandwidth VCSEL devices. In: LEOS 2006, 19th annual meeting of the IEEE lasers and electro-optics society, Montreal, Que., Canada, TuBB1 (invited), October 2006, pp 400–401
16. Hopfer F, Mutig A, Strittmatter A, Fiol G, Moser P, Bimberg D, Shchukin VA, Ledentsov NN, Lott JA, Quast H, Kuntz M, Mikhrin SS, Krestnikov IL, Livshits DA, Kovsh AR, Bornholdt C (2008) High-speed directly and indirectly modulated VCSELs. In: IEEE lasers and electro-optics society, IEEE electron devices society, 20th international conference on indium phosphide and related materials, IPRM 2008, Versailles, France, 25–29 May 2008
17. Bowers JE, Chen H-W, Liang D, Oakley DC, Napoleone A, Chapman DC, Chen C-L, Juodawlkis PW (2009) Integration using the hybrid silicon platform. Invited paper 1.7.1, European conference on optical communication (ECOC), Vienna, Austria, 20–24 September 2009
18. Lott JA (2002) Fabrication and applications of lift-off vertical cavity surface emitting laser (VCSEL) disks. Invited talk, Proceedings SPIE 4649-31, Photonics West, 20–25 January 2002, San Jose, CA, USA, 4 June 2002, pp 203–210
19. Raley JA, Lott JA, Nelson TR Jr, Stintz A, Malloy KJ (2002) Interconnected lift-off VCSELs for microcavity device arrays. In: Proceedings IEEE LEOS summer topicals: VCSEL and microcavity lasers, Mont Tremblant, Canada, 15–17 July 2002, pp MH2-11–MH2-12, 7 November 2002
20. Hill MT, Oei Y-S, Smalbrugge B, Zhu Y, de Vries T, van Veldhoven PJ, van Otten FWM, Eijkemans TJ, Turkiewicz JP, de Waardt H, Jan Geluk E, Kwon S-H, Lee Y-H, Nötzel R, Smit MK (2007) Lasing in metallic-coated nanocavities. Nature Photonics 1: 589–594, 16 September 2007
21. Altug H, Englund D, Vuckovic J (2006) Ultrafast photonic crystal nanocavity laser. Nat Phys 2:484–488
22. Chang S-W, Chuang SL (2008) Plasmonic nano-laser based on metallic bowtie cavity. In: Conference paper, quantum electronics and laser science conference (QELS), San Jose, CA, Nanoplasmonics IV (QTuJ), QTuJ5, 4 May 2008
23. Chang S-W, Ni C-YA, Chuang SL (2008) Theory for bowtie plasmonic nanolasers. Opt Express 16(14):10580–10595

Appendix

Appendix A: Measurement Setup

For measurements of the CW laser characteristics setups shown in Fig. A.1 and in Fig. A.2 were used. For L-U-I measurements the VCSEL was driven by the digital sourcemeter Keithley 2400-LV and the output light was collected by the calibrated integration sphere, as demonstrated in Fig. A.1. The photocurrent from the integration sphere was measured by the second sourcemeter. Both sourcemeters were controlled by a computer utilizing a LabView program. The temperature of the VCSEL was precisely controlled by a vacuum thermochuck Temptronic TP03010B, which can set the chuck temperature to values from 0 up to 200°C. In all measurements described in this work exclusively the digital sourcemeters Keithley 2400-LV were used.

To measure emission spectra the VCSEL was driven by a sourcemeter and the output light was coupled into a graded index multimode fiber with 50 μm or 62.5 core diameter connected to an optical spectrum analyzer ANDO AQ6317C or HP 70951B, as shown in Fig. A.2. In all measurements with optical fibers described in this work exclusive graded index multimode fibers with 50 or 62.5 μm core diameter were used. The temperature of the VCSEL was controlled by the vacuum thermochuck.

In Fig. A.3 measurement setup for the small signal modulation response measurements is presented. These measurements are the key technique to get access to the internal device parameters and therefore should be carried out very carefully. The operation current was determined by a digital sourcemeter, which was connected via network analyzer HP 8722C to the laser. The constant signal from the sourcemeter was overlapped with a small harmonic signal from the network analyzer and delivered to the device under test. The output optical signal from the VCSEL was coupled into a fiber and guided to a calibrated photodetector or a calibrated photoreceiver, which was connected to the second port of the network analyzer. For the measurements of the VCSELs emitting around 850 nm the photodetector VIS D30-850M was used. Measurements of the VCSELs

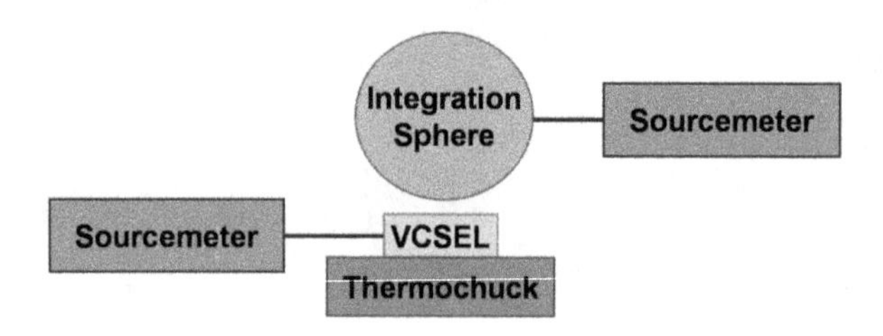

Fig. A.1 Setup for L-U-I measurements

emitting around 980 nm were carried out using the photoreceiver DSC-R401HG, the photoreceiver New Focus 1554-B-50 or the photodetector New Focus 1434-50-M. Measurements of the scattering parameters S_{11} and S_{21} were carried out at different chuck temperatures.

The measurement setup for optical eye diagrams is shown in Fig. A.4. The constant signal of the digital sourcemeter is combined with the amplified non-return to zero (NRZ) bit pattern with 2^7-1 pseudo-random bit sequence (PRBS) generated by the bit pattern generator SHF 12100B. Thereby electrical amplifiers from SHF were used. The combined signal is guided to the VCSEL under test and the output power is coupled via a multimode fiber with 2–4 m length (back-to-back (BTB) configuration) into a photodetector or photoreceiver. After the photodetector one or two electrical amplifiers were used to amplify the signal, which was then recorded by the digital oscilloscope Agilent DCA-J 86100C.

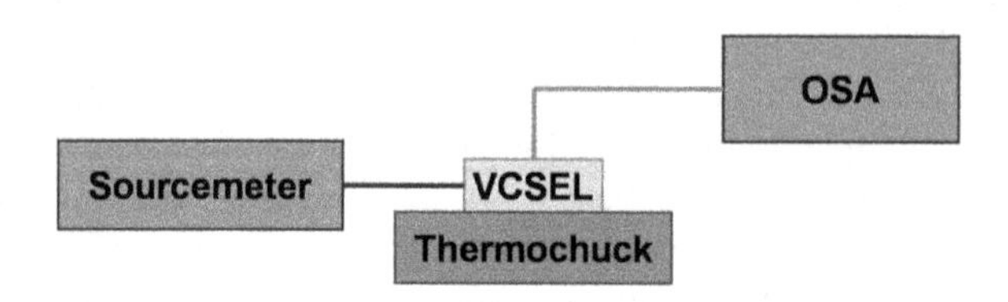

Fig. A.2 Setup for measurements of optical emission spectra

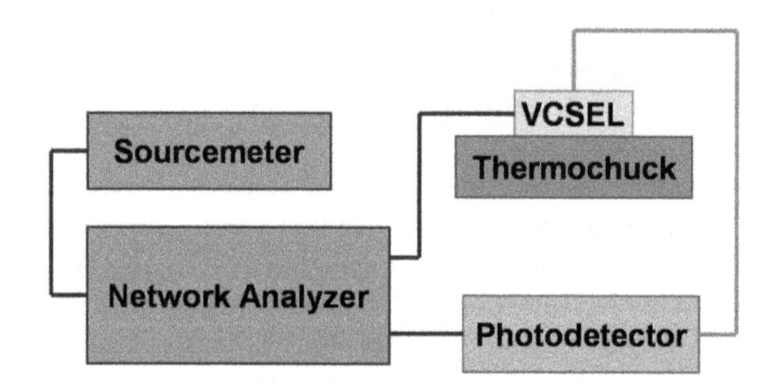

Fig. A.3 Setup for small signal modulation measurements

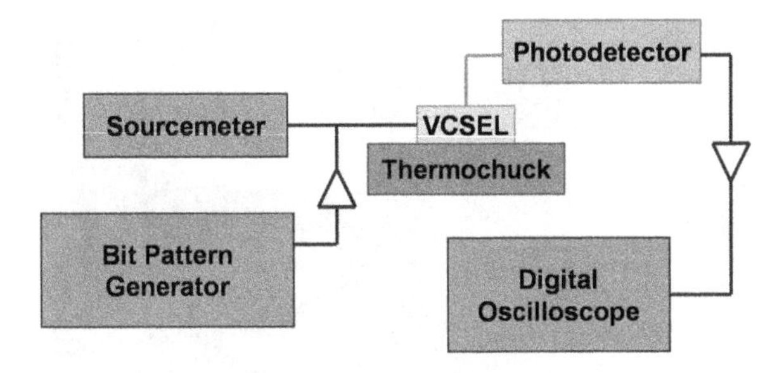

Fig. A.4 Setup for measurements of optical eye diagrams

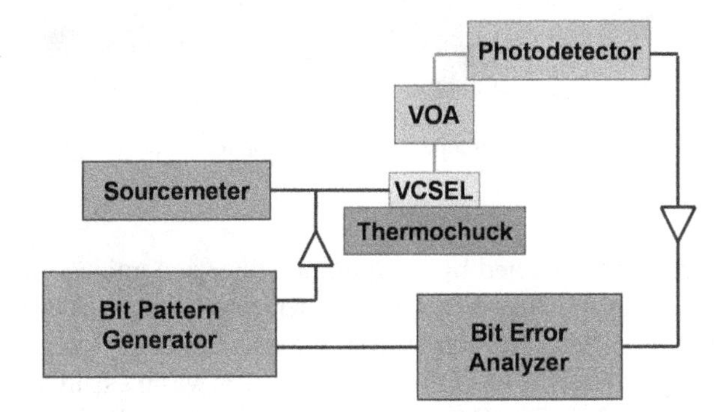

Fig. A.5 Setup for measurements of bit error rates

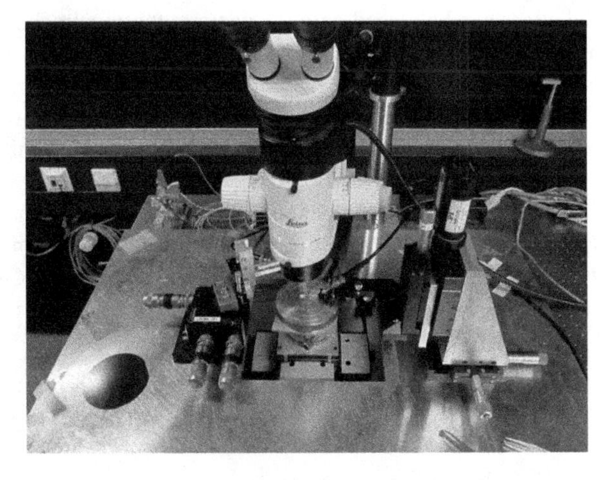

Fig. A.6 Picture of the measuring station with the microscope and precision stages

The VCSEL temperature was controlled by the vacuum thermochuck.

To measure the bit error rates the VCSEL was driven by a combined signal consisted of the constant part and the amplified bit pattern, as shown in Fig. A.5.

Fig. A.7 Picture of the digital oscilloscope, signal generator and bit pattern generator with the bit error analyzer

The output signal was collected by a multimode fiber and guided via the variable optical attenuator JDSU OLA-54 to a photodetector with an electrical amplifier after it or a photoreceiver. The signal from the photodetector was analyzed using the bit error analyzer SHF 11100B. The optical power was measured by the optical powermeter JDSU OLP-55. The temperature was controlled by the vacuum thermochuck.

In the Fig. A.6 a picture of the on-wafer measuring station is shown. The wafer with fabricated devices can be placed onto the vacuum thermochuck mounted on a 2-axis precision stage. The integration sphere or a fiber holder is mounted on the 3-axis precision stage left and the high frequency prober is placed onto the 3-asix precision stage right. The positions of the VCSEL under test, the prober and the fiber can be controlled via the microscope. The whole setup is built up on the stabilized table to avoid mechanical vibrations.

Fig. A.7 shows a picture of the rack on the left side with the bit pattern generator SHF 12111B, bit error analyzer SHF 11100B and signal generator Agilent E8247C used as a clock. The digital oscilloscope Agilent DCA-J 86100C is shown on the right side as well.

Curriculum Vitae

Alex Mutig

Personal
Birthday: October 20th, 1978
Birth place: Semipalatinsk, Kazakhstan

School and High School
1985–1993: 9 years middle school, Semipalatinsk, Kazakhstan
1993–1995: 10 and 11 year at the Pedagogical Institute of Semipalatinsk, Semipalatinsk, Kazakhstan

University
September 1995–July 1999: Study of physics at the University "Semej", Semipalatinsk, Kazakhstan
October 1999: Move to Germany
February 2000–March 2000: Language course (German) at GFBM, Berlin, Germany
October 2000–May 2004: Study of physics at the Technical University of Berlin, Berlin, Germany

Dissertation
June 2004–July 2010: PhD student in the group of Prof. Dr. D. Bimberg at the Technical University of Berlin, Berlin, Germany

List of Publications and Conference Presentations

Journal Papers

1. **A. Mutig,** S. A. Blokhin, A. M. Nadtochiy, G. Fiol, J. A. Lott, V. A. Shchukin, N. N. Ledentsov and D. Bimberg, “Frequency response of large aperture oxide-confined 850 nm vertical cavity surface emitting lasers,” Applied Physics Letters, Vol. 95, 131101, 2009
2. **A. Mutig,** G. Fiol, K. Pötschke, P. Moser, D. Arsenijevic, V. A. Shchukin, N. N. Ledentsov, S. S. Mikhrin, I. L. Krestnikov, D. A. Livshits, A. R. Kovsh, F. Hopfer and D. Bimberg, “Temperature-dependent small-signal analysis of high-speed high-temperature stable 980-nm VCSELs,” IEEE Journal of Selected Topics in Quantum Electronics, Vol. 15, No. 3, pp. 679–686, May/June 2009
3. **A. Mutig,** G. Fiol, P. Moser, D. Arsenijevic, V. A. Shchukin, N. N. Ledentsov, S. S. Mikhrin, I. L. Krestnikov, D. A. Livshits, A. R. Kovsh, F. Hopfer and D. Bimberg, “120°C 20 Gbit/s operation of 980 nm VCSEL,” Electronics Letters, Vol. 44, No. 22, 23rd October 2008
4. S. A. Blokhin, J. A. Lott, **A. Mutig,** G. Fiol, N. N. Ledentsov, M. V. Maximov, A. M. Nadtochiy, V. A. Shchukin and D. Bimberg, “Oxide-confined 850 nm VCSELs operating at bit rate up to 40 Gbit/s,” Electronics Letters, Vol. 45, No. 10, 7th May 2009
5. F. Hopfer, **A. Mutig,** G. Fiol, M. Kuntz, V. A. Shchukin, V. A. Haisler, T. warming, E. Stock, S. S. Mikhrin, I. L. Krestnikov, D. A. Livshits, A. R. Kovsh, C. Bornholdt, A. Lenz, H. Eisele, M. Dähne, N. N. Ledentsov and D. Bimberg, “20 Gb/s 85°C error-free operation of VCSELs based on submonolayer deposition of quantum dots,” IEEE Journal of Selected Topics in Quantum Electronics, Vol. 13, No. 5, pp. 1302–1308, September/October 2007
6. F. Hopfer, **A. Mutig,** M. Kuntz, G. Fiol, N. N. Ledentsov, V. A. Shchukin, S. S. Mikhrin, D. L. Livshits, I. L. Krestnikov, A. R. Kovsh, N. D. Zakharov, P. Werner and D. Bimberg, “Single-mode submonolayer quantum-dot vertical-cavity surface-emitting lasers with high modulation bandwidth,” Applied Physics Letters, Vol. 89, 141106, 2006
7. F. Hopfer, I. Kaiander, A. Lochmann, **A. Mutig,** S. Bognar, M. Kuntz, U. W. Pohl, V. A. Haisler and D. Bimberg, “Vertical-cavity surface-emitting quantum-dot laser with low threshold current grown by metal-organic vapor phase epitaxy,” Applied Physics Letters, Vol. 89, 061105, 2006
8. L. Olejniczak, M. Sciamanna, H. Thienpont, K. Panajotov, **A. Mutig,** F. Hopfer and D. Bimberg, “Polarization switching in quantum-dot vertical-cavity surface-emitting lasers,” IEEE Photonics Technology Letters, Vol. 21, No. 14, pp. 1008–1010, July 15, 2009
9. M. Laemmlin, G. Fiol, M. Kuntz, F. Hopfer, **A. Mutig,** N. N. Ledentsov, A. R. Kovsh, C. Schubert, A. Jacob, A. Umbach and D. Bimberg, “Quantum dot based devices at 1.3 μm: direct modulation, mode-locking and VCSELs,” Phys. Stat. Sol. (c) **3,** No. 3, 391–394, 2006

10. S. A. Blokhin, L. Ya. Karachinsky, I. I. Novikov, S. M. Kuznetsov, N. Yu. Gordeev, Y. M. Shernyakov, A. V. Savelyev, M. V. Maximov, **A. Mutig,** F. Hopfer, A. R. Kovsh, S. S. Mikhrin, I. L. Krestnikov, D. A. Livshits, V. M. Ustinov, V. A. Shchukin, N. N. Ledentsov and D. Bimberg, "MBE-grown ultra-large aperture single-mode vertical-cavity surface-emitting laser with all-epitaxial filter section," Journal of Crystal Growth, 301–302, pp. 945–950, 2007
11. W. Hofmann, M. Müller, A. Nadtochiy, C. Meltzer, **A. Mutig,** G. Böhm, J. Rosskopf, D. Bimberg, M.-C. Amann and C. Chang-Hasnain, "22-Gb/s long wavelength VCSELs," Optics Express, Vol. 17, No. 20, pp. 17547–17554, 28 September 2009
12. N. N. Ledentsov, D. Bimberg, F. Hopfer, **A. Mutig,** V. A. Shchukin, A. V. Savelyev, G. Fiol, E. Stock, H. Eisele, M. Dähne, D. Gerthsen, U. Fischer, D. Litvinov, A. Rosenauer, S. S. Mikhrin, A. R. Kovsh, N. D. Zakharov, P. Werner, "Submonolayer quantum dots for high speed surface emitting lasers," Nanoscale Res. Lett., 2:417–429, 2007

Conference Presentations

1. **A. Mutig,** S. Blokhin, A. M. Nadtochiy, G. Fiol, J. A. Lott, V. A. Shchukin, N. N. Ledentsov, D. Bimberg, "High-speed 850 nm oxide-confined VCSELs for DATACOM applications," Vertical-Cavity Surface-Emitting Lasers XIV, Photonics West 2010, 23–28 January 2010, San Francisco, California, USA, Proceedings of SPIE, Vol. 7615, 76150 N, Invited Paper, 5th February 2010
2. **A. Mutig,** J. Lott, S. Blokhin, G. Fiol, A. Nadtochiy, V. Shchukin, N. Ledentsov and D. Bimberg, "High speed VCSELs for short reach DATACOM applications," Spring Meeting of the German Physical Society (DPG), 21–26 March 2010, Regensburg, Germany, DS2.6, Topical Talk, 2010
3. **A. Mutig,** G. Fiol, P. Moser, F. Hopfer, M. Kuntz, V. A. Shchukin, N. N. Ledentsov, S. S. Mikhrin, I. L. Krestnikov, D. A. Livshits, A. R. Kovsh, D. Bimberg, "120°C 20 Gbit/s operation of 980 nm single mode VCSEL," IEEE 21st International Semiconductor Laser Conference (ISLC), 14–18 September 2008, Sorrento, Italy, Paper MB2, Conference Digest, pp. 9–10, 30th September 2008
4. **A. Mutig,** F. Hopfer, G. Fiol, M. Kuntz, V. Shchukin, N. N. Ledentsov, S. S. Mikhrin, I. L. Krestnikov, D. A. Livshits, A. R. Kovsh, C. Bornholdt, D. Bimberg, "12.5 Gbit/s 1250 nm VCSELs based on low-temperature grown quantum dots," European Semiconductor Laser Workshop (ESLW), 14–15 September 2007, Berlin, Germany, A4, 2007
5. N. N. Ledentsov, J. A. Lott, V. A. Shchukin, D. Bimberg, **A. Mutig,** T. D. Germann, J.-R. Kropp, L. Y. Karachinsky, S. A. Blokhin, A. M. Nadtochiy, "Optical components for very short reach applications at 40 Gb/s and beyond," Physics and Simulation of Optoelectronic Devices XVIII, Photonics West, 23–28 January 2010, San Francisco, California, USA, Proceedings of SPIE, Vol. 7597, 75971F, 25th February 2010

6. D. Bimberg, S. A. Blokhin, **A. Mutig,** A. M. Nadtochiy, G. Fiol, P. Moser, D. Arsenijevic, F. Hopfer, V. A. Shchukin, J. A. Lott and N. N. Ledentsov, "Nano-VCSELs for the terabus," 17th International Symposium "Nanostructures: Physics and Technology", 22–26 June 2009, Minsk, Belarus, OPS.01pl, 2009
7. J. A. Lott, V. A. Shchukin, N. N. Ledentsov, A. Stinz, F. Hopfer, **A. Mutig,** G. Fiol, D. Bimberg, S. A. Blokhin, L. Y. Karachinsky, I. I. Novikov, M. V. Maximov, N. D. Zakharov, P. Werner, "20 Gbit/s error free transmission with ~850 nm GaAs-based vertical cavity surface emitting lasers (VCSELs) containing InAs–GaAs submonolayer quantum dot insertions," Physics and Simulation of Optoelectronic Devices XVII, Photonics West 2009, 24–29 January 2009, San Jose, California, USA, Proceedings of SPIE, Vol. 7211, 721114, 24th February 2009
8. N. N. Ledentsov, J. A. Lott, V. A. Shchukin, H. Quast, F. Hopfer, G. Fiol, **A. Mutig,** P. Moser, T. Germann, A. Strittmatter, L. Y. Karachinsky, S. A. Blokhin, I. I. Novikov, A. M. Nadtochiy, N. D. Zakharov, P. Werner, D. Bimberg, "Quantum dot insertions in VCSELs from 840 to 1300 nm: growth, characterization, and device performance," Quantum Dots, Particles, and Nanoclusters VI, Photonics West 2009, 24–29 January 2009, San Jose, California, USA, Proceedings of SPIE, Vol. 7224, 72240P, 17th February 2009
9. F. Hopfer, **A. Mutig,** G. Fiol, P. Moser, D. Arsenijevic, V. A. Shchukin, N. N. Ledentsov, S. S. Mikhrin, I. L. Krestnikov, D. A. Livshits, A. R. Kovsh, M. Kuntz, D. Bimberg, "120°C 20 Gbit/s operation of 980 nm VCSEL based on sub-monolayer growth," Vertical-Cavity Surface-Emitting Lasers XIII, Photonics West 2009, 24–29 January 2009, San Jose, California, USA, Proceedings of SPIE, Vol. 7229, 72290C, 6th February 2009
10. F. Hopfer, **A. Mutig,** A. Strittmatter, G. Fiol, P. Moser, V. A. Shchukin, N. N. Ledentsov, J. A. Lott, H. Quast, M. Kuntz, S. S. Mikhrin, I. L. Krestnikov, D. A. Livshits, A. R. Kovsh, C. Bornholdt, D. Bimberg, " High-speed directly and indirectly modulated VCSELs," 20th International Conference on Indium Phosphide and Related Materials (IPRM), 25–29 May 2008, Versailles, France, pp. 1–6, 9th December 2008
11. V. A. Shchukin, N. N. Ledentsov, J. A. Lott, H. Quast, F. Hopfer, L. Ya. Karachinsky, M. Kuntz, P. Moser, **A. Mutig,** A. Strittmatter, V. P. Kalosha, D. Bimberg, "Ultra high-speed electro-optically modulated VCSELs: modeling and experimental results," Physics and Simulation of Optoelectronic Devices XVI, Photonics West 2008, 19–24 January 2008, San Jose, California, USA, Proceedings of SPIE, Vol. 6889, 68890H, 22nd February 2008
12. F. Hopfer, **A. Mutig,** G. Fiol, M. Kuntz, V. Shchukin, N. N. Ledentsov, S. S. Mikhrin, I. L. Krestnikov, D. A. Livshits, A. R. Kovsh and D. Bimberg, "High speed 1225 and 1250 VCSELs based on low-temperature grown quantum dots," The European Conference on Lasers and Electro-Optics and the International Quantum Electronics Conference (CLEO Europe/IQEC), 17–22 June 2007, Munich, Germany, 2007

13. F. Hopfer, **A. Mutig,** G. Fiol, M. Kuntz, V. Shchukin, N. N. Ledentsov, S. S. Mikhrin, I. L. Krestnikov, D. A. Livshits, A. R. Kovsh, C. Bornholdt, D. Bimberg, "10 Gbit/s 1250 nm VCSELs based on low-temperature grown quantum dots," International Workshop on Physics and Applications of Semiconductor Lasers (PHASE), 28–30 March 2007, Metz, France, 29th March, Session 5, 12:20, 2007
14. F. Hopfer, **A. Mutig,** G. Fiol, M. Kuntz, V. Shchukin, N. N. Ledentsov, S. S. Mikhrin, I. L. Krestnikov, D. A. Livshits, A. R. Kovsh, C. Bornholdt, D. Bimberg, "1250 nm high speed VCSELs based on low-temperature grown quantum dots," COST 288, 26–27 March 2007, Metz, France, 26th March, 11:20, 2007
15. N. N. Ledentsov, F. Hopfer, **A. Mutig,** V. A. Shchukin, A. V. Savelyev, G. Fiol, M. Kuntz, V. A. Haisler, T. Warming, E. Stock, S. S. Mikhrin, A. R. Kovsh, C. Bornholdt, H. Eisele, M. Dähne, N. D. Zakharov, P. Werner, D. Bimberg, " Novel concepts for ultrahigh-speed quantum-dot VCSELs and edge-emitters," Physics and Simulation of Optoelectronic Devices XV, Photonics West 2007, 20–25 January 2007, San Jose, California, USA, Proceedings of SPIE, Vol. 6468, 64681O, 7th February 2007
16. F. Hopfer, **A. Mutig,** G. Fiol, M. Kuntz, S. S. Mikhrin, I. L. Krestnikov, D. A. Livshits, A. R. Kovsh, V. Shchukin, N. N. Ledentsov, C. Bornholdt, N. D. Zakharov, P. Werner, D. Bimberg, " Ultra high speed submonolayer quantum-dot vertical-cavity surface-emitting lasers," Sixth IEEE Conference on Nanotechnology (IEEE-NANO), 16–20 July 2006, Cincinnati, Ohio, USA, Vol. 2, pp. 749–751, 30th October 2006
17. F. Hopfer, **A. Mutig,** G. Fiol, M. Kuntz, V. Shchukin, N. N. Ledentsov, S. S. Mikhrin, I. L. Krestnikov, D. A. Livshits, A. R. Kovsh, C. Bornholdt, D. Bimberg, "20 Gb/s 85°C Error free operation of VCSEL based on submonolayer deposition of quantum dots," IEEE 20th International Semiconductor Laser Conference (ISLC), 17–21 September 2006, Kohala Coast, Big Island of Hawaii, Hawaii, USA, Conference Digest, Paper WC3, pp. 119–120, 9th October 2006
18. F. Hopfer, **A. Mutig,** M. Kuntz, G. Fiol, D. Bimberg, N. N. Ledentsov, V. A. Shchukin, S. S. Mikhrin, D. A. Livshits, I. L. Krestnikov, A. R. Kovsh and C. Bornholdt, "Ultra high speed submonolayer quantum dot vertical-cavity surface-emitting lasers," International Conference on the Physics of Semiconductors (ICPS), 24–28 July 2006, Vienna, Austria, TuM1b.4, 2006
19. F. Hopfer, **A. Mutig,** G. Fiol, M. Kuntz, S. S. Mikhrin, I. L. Krestnikov, D. A. Livshits, A. R. Kovsh, C. Bornholdt, V. Shchukin, N. N. Ledentsov, V. Gaysler, N. D. Zakharov, P. Werner, D. Bimberg, " 20-Gb/s direct modulation of 980 nm VCSELs based on submonolayer deposition of quantum dots," Workshop on Optical Components for Broadband Communication, 27th July 2006, Stockholm, Sweden, Proceedings of SPIE, Vol. 6350, 635003, 6th July 2006
20. F. Hopfer, **A. Mutig,** G. Fiol, M. Kuntz, S. S. Mikhrin, I. L. Krestnikov, D. A. Livshits, V. A. Shchukin, A. R. Kovsh, N. N. Ledentsov, C. Bornholdt,

D. Bimberg, "High speed performance of 980 nm VCSELs based on submonolayer quantum dots," The Conference on Lasers and Electro-Optics and the International Quantum Electronics Conference (CLEO/QELS), 21–26 May 2006, Long Beach, California, USA, Techn. Digest of CLEO/QELS and PhAST 2006, The Optical Society of America CD, CPDB2, 2006

21. F. Hopfer, M. Kuntz, M. Lämmlin, N. N. Ledentsov, A. R. Kovsh, S. S. Mikhrin, I. Kaiander, V. Haisler, A. Lochmann, **A. Mutig,** C. Schubert, N. Grote, A. Umbach, V. M. Ustinov, U. W. Pohl, D. Bimberg, "Quantum dot photonics: edge emitter, amplifier and VCSEL," Second International Conference on Advanced Optoelectronics and Lasers (CAOL), 12–17 September 2005, Yalta, Crimea, Ukraine, Invited paper, Proceedings of CAOL 2005, Vol. 1, pp.1–4, 27th December 2005

CPSIA information can be obtained
at www.ICGtesting.com
Printed in the USA
LVHW051925291219
641987LV00002B/100/P

9 783642 165696